Marxism in the Age of Ecological Catastrophe

Marxism in the Age of Ecological Catastrophe offers a stimulating discussion on the ecological unfeasibility of capitalist society.

Divided into three parts, Eduardo Sá Barreto begins by providing a reconstruction of Marx's theory of value and articulating it into an ecological critique of this society. Part II surveys key debates between some of today's most representative Marxist ecologists. Part III explores political approaches, tactical and strategic issues to see whether they align with the gravity of the challenges facing humanity or not.

Located at the intersection of the natural and social sciences, *Marxism in the Age of Ecological Catastrophe* will be of interest to scholars of political science, economics, ecology, climatology, demography, geography, and sociology.

Eduardo Sá Barreto is an associate professor of economics at the Federal Fluminense University (UFF), Brazil, and a researcher at the Interdisciplinary Center for Studies and Research on Marx and Marxism (NIEP-Marx).

Marx and Marxisms: New Horizons

Edited by Marcello Musto

The peer-reviewed series *Marx and Marxisms: New Horizons* (edited by Marcello Musto, with Francesca Antonini as Assistant Editors) comprises rigorous scholarly books, accessible to general readers, that offer innovative and critical works in the field of Marx studies and Marxisms. The series publishes monographs, edited collections, and anthologies, by both prestigious and emerging international experts, in the fields of political theory, history of political thought, sociology, political philosophy, and heterodox economics. The books in this series provide original investigations within the Marxist tradition, push the boundaries of accepted interpretations and existing literatures, bring different concepts and thinkers into new relationships, and inspire significant conversations for today. They come from a wide range of academic disciplines, subject matters, political perspectives, cultural backgrounds, and geographical areas, producing an eclectic and informative collection that appeals to a diverse and international audience.

Walter Benjamin and Antonio Gramsci
A Missed Encounter
Edited by Dario Gentili, Elettra Stimilli, Gabriele Guerra

Can Common People Govern?
Political Parties, Movements, and Uprisings
Jacques Bidet

Marxism in the Age of Ecological Catastrophe
Theory and Praxis
Eduardo Sá Barreto

For more information about this series, please visit: www.routledge.com/Marx-and-Marxisms/book-series/MM

Marxism in the Age of Ecological Catastrophe

Theory and Praxis

Eduardo Sá Barreto

Translation from the Portuguese *Ecologia marxista para pessoas sem tempo,* by Débora Cunha

Routledge
Taylor & Francis Group
NEW YORK AND LONDON

First published 2025
by Routledge
605 Third Avenue, New York, NY 10158

and by Routledge
4 Park Square, Milton Park, Abingdon, Oxon, OX14 4RN

Routledge is an imprint of the Taylor & Francis Group, an informa business

Translation from the Portuguese Ecologia marxista para pessoas sem tempo, by Débora Cunha

ISBN: 978-1-032-84392-6 (hbk)
ISBN: 978-1-032-84393-3 (pbk)
ISBN: 978-1-003-51251-6 (ebk)

DOI: 10.4324/9781003512516

Typeset in Times New Roman
by Newgen Publishing UK

This book is dedicated to João Leonardo Medeiros and Marcelo Badaró Mattos.

Contents

Foreword

The material you now have in your hands is the consolidation of a course I have been teaching for more than ten years, to undergraduate and graduate students. Evidently, a lot of new literature has been added over the years and some topics have entered the scope of the course, theoretical demonstrations have been refined, and the level of abstraction of the considerations has become closer and closer to the concrete, as contemporary issues have been incorporated.

I'm sure there would be a lot to advance and deepen. Topics such as ecological imperialism, environmental racism, and just transition are not covered, for example. At the same time, the theoretical reasoning does not sufficiently advance on topics such as structural crisis, fictitious capital, and precarious labour in their relation to ecological issues. Occasionally, ecosocialist thinkers are only partially presented, without their main virtues or weaknesses being fully discussed (or presented at all).

Despite these shortcomings, however, the reader may come to realise that filling in all the gaps is unattainable and, most importantly, unnecessary. The reason is twofold and, I believe, relevant. Firstly, Marxist ecology has already produced (even if not finished) a demolishing ecological critique of capitalism. With the critique we have at our disposal, which I present in the opening chapters of this book, it is possible to demonstrate the necessarily devastating nature of the logic of capital over the planet. Therefore, with the critique we already have, there can no longer be any doubt that capitalism needs to be overcome. Theoretical, thematic, or bibliographical additions to this critique will further strengthen its power, but as it stands, it razes the matter to the ground.

Secondly, this same ecological critique, when equipped with the best scientific understanding we have of the processes of planetary destabilisation/destruction and the vast body of evidence about these processes, leads us to conclude that overcoming capitalist society is not only necessary, but also urgent. At the moment, humanity is a sort of runaway car about to hit a concrete wall at very high speed. It does not make much difference whether the collision is at 350 km/h, 375 km/h, 400 km/h, or whether we're in the driver's seat, the passenger's seat, or the back seat. Whether we're wearing a seatbelt or a helmet does not make that much difference. It is enough to know that the speed implies a deadly collision and that humanity needs to find ways to pull the emergency brake, make an unprecedented U-turn, and survive the rollover.

From now on, these are the questions my research will focus on. Not to contribute to improving well-known "small-wining" or "time-buying" techniques. On the contrary, if the ecological critique of capitalism tells us that a revolution is necessary and urgent, we must find ways to achieve it.

Rio de Janeiro, June 26, 2024

Acknowledgements

The translation of the manuscript for this edition received financial support from Capes Proex of the Postgraduate Program in Economics at the Universidade Federal Fluminense.

1 Introduction

Hegemonic environmental thinking will not do

There are two major schools of thought regarding environmental issues in economics: Environmental Economics and Ecological Economics. Environmental Economics, the oldest, descends from the type of economic theory developed from the last quarter of the 19th century onwards and became known at the beginning of the 20th century as the neoclassical school. As will be further explored, its emphasis is on the ability of markets to self-adjust and on the policies capable of generating the appropriate incentives for this adjustment to take place in an environmentally sustainable way. Ecological Economics, which is more interdisciplinary and initially less dependent on neoclassical theories, emerged as a critique of the pillars of Environmental Economics.

Therefore, its pillars are always presented as counterpoints to those of Environmental Economics. For this reason, it is helpful to present the founding ideas of Ecological Economics now.[1] In them, we can envision decisive aspects of both schools. The fundamental topics to be covered are (i) the notion of metabolism, (ii) the importance of thermodynamics, (iii) the perspective on the production process, (iv) the perspective on technology, and (v) the perspective on economic growth.

Before diving into the specifics, it is worth delineating three critical concepts from physics: isolated systems, closed systems, and open systems. Isolated systems do not exchange energy or matter with their surroundings. In other words, there is no input or output of matter or energy in this system. Closed systems have energy input and output but no matter input or output. Open systems regularly exchange matter and energy with their surroundings.[2]

How is this distinction relevant to the issue discussed in this chapter? Environmental and Ecological Economics have entirely different views on the relationship between the economic sphere and nature. Roughly speaking, for Environmental Economics, the economic system is a closed system. Nature takes the role of a mere subsystem of the economic system. On the other hand, for Ecological Economics, planet Earth is a closed system, and the economy is just an open subsystem.

DOI: 10.4324/9781003512516-1

This difference is crucial. In one tradition, nature is subordinate to the economy, while in the other, the economy is subordinate to the natural world. One way to better understand this is by exploring the first of the five topics listed above: metabolism.

Most of those who have studied economics are familiar with the circular flow diagram. It shows two flows: a material flow, originated and moved by families and companies, and a monetary flow, originated and moved by markets for goods and services on one side and factors of production on the other. It is not essential to understand this as an economist would. The key is to realise that this circular flow is self-sufficient.

This idea is patently fragile, especially when presented in this hyper-schematic way. Even so, it is part of an economist's education since its very beginning. It should, therefore, come as no surprise that Environmental Economics, a descendant of the neoclassical tradition, transposes analogous reasoning to environmental issues. In the circular flow, nothing comes in, and nothing gets out. There is no input of energy or matter. There is no outflow or release of waste and energy. The circular flow is the fundamental characterisation of the economic sphere as an isolated system. It is almost impossible to draw any analogy between the idea of metabolism and a representation of how the system works such as this one.

Ecological Economics bases its criticism of this notion on the second law of thermodynamics, which asserts that energy can never be fully converted into useful work.[3] This means that every form of energy consumption must always generate a residue; a fraction of it is always dissipated as heat in the transformation process. The topic will be further examined later in this book. However, to understand the Ecological Economics perspective on metabolism, we already need this insight that movement and transformation always generate a certain degree of deterioration.

The way Ecological Economics originally articulates thermodynamics and metabolism goes back to the Romanian thinker Georgescu-Roegen. This author and Ecological Economics have taken some creative liberties with the concept of entropy, borrowed from thermodynamics, a problem analysed in the next topic. For now, it is enough to realise that low-entropy matter and energy are highly functional for useful purposes. On the other hand, high-entropy matter and energy have a low possibility of useful application because of higher levels of degradation. Georgescu-Roegen applies these notions to say that all economic life feeds on low-entropy matter and energy and generates high-entropy waste as a by-product.[4]

From this flow of transformation of low-entropy matter–energy into high-entropy matter–energy comes the concept of *socio-environmental metabolic flow*, which seeks to grasp the multiple relationships that human societies establish with nature. In other words, it seeks to consider the existence of human beings not simply as economic beings but also as natural and physical beings.

There is a pronounced difference between this perspective on metabolism and the notion of circular flow mentioned earlier. In this conception proposed by Ecological Economics, socio-environmental metabolism is regulated both by natural laws governing physical processes and by institutionalised norms governing the division of labour, the distribution of wealth, consumption patterns, etc.[5]

Before advancing, it is worth mentioning that the first chapter in the second part of this book returns to a discussion of metabolism. However, one should be careful not to conflate these two approaches. Although there are possible parallels, they are two quite different lines of reasoning.

Georgescu-Roegen's way of proposing the notion of metabolism brings us to the second of the five points: thermodynamics. Once again, it is helpful to underline the opposition between Environmental and Ecological Economics. In this specific case, the opposition is between the influence of classical mechanics on Environmental Economics and the influence of thermodynamics on Ecological Economics.

According to the critique produced by Ecological Economics, Environmental Economics borrows three fundamental ideas from mechanics. First, that total usable energy is constant. In other words, there is no movement from low-entropy energy to high-entropy energy; there is no waste. Second, predictability. It is assumed that predicting future developments with relative precision is always possible. In other words, uncertainty has little (if any) room in this theoretical framework. Third, reversibility. There are no points of no return. It is assumed that returning to previous states is always possible.

Curiously, this mechanistic view, so to speak, will also find at least one point of support, albeit a very fragile one, in thermodynamics. That is because Environmental Economics considers the economy an isolated system and the first law of thermodynamics (the law of conservation) states that, in isolated systems, the amount of energy remains constant; there is no creation or destruction of energy, only transformation.[6]

However, when moving on to the second law, we definitely leave the scope of Environmental Economics and enter the Ecological Economics territory. The second law tells us that in an isolated system, the quality of energy tends to degrade over time. For our purposes, degrading means becoming unavailable to perform useful work. Transformations in energy make it progressively unusable. In concrete terms, every energy transformation involves using a fraction of it to perform work and the dissipation of the remaining fraction in the form of heat.[7]

The mechanistic conceptual apparatus of Environmental Economics is not able to comprehend this unidirectional movement of energy. The notion of reversibility is incompatible with the second law. Transposed into the reasoning of Ecological Economics, it informs us that each economic process carries a good deal of irreversibility, including its interaction with the natural environment.

This leads to the conclusion that economic activity *always* generates waste, whether material or energy.

Although this is an obviously correct conclusion, it is now necessary to briefly comment on two creative liberties in the way Ecological Economics theoretically resorts to the second law of thermodynamics. First, the law refers to an isolated system. Planet Earth is not an isolated system but a closed one. Indeed, everyone knows that a permanent influx of energy is coming from the Sun. This is not a mere detail but a decisive factor when considering entropy and usable energy. Second, the law only refers to energy. It is true that, at the beginning of the 20th century, Einstein had already discovered the equivalence between mass and energy, presenting the famous formula $E = mc^2$. However, in the context of economic life, in which everyday speeds are extraordinarily far from the speed of light, extrapolating entropy to matter, as Ecological Economics does, is problematic, to say the least.

Having made that warning, and keeping this reasoning in mind, the third of the five points to be analysed is the perspective on the production process. One of the fundamental foundations of the neoclassical theoretical framework on which Environmental Economics is built is the idea that factors of production can be substituted for each other. Also, that the composition of factors used in the production process will depend on their relative prices and the rational calculation of the economic agent, who seeks their most efficient combination, aiming to maximise profits. According to the critique placed by Ecological Economics, the principle of substitutability between factors of production prevents any consideration about qualitative differences between the elements that take part in the production process. The only challenge posed by this principle is finding the most efficient allocation of the available factors.

If all factors are substitutable, they can be taken as being of a similar nature. In this sense, natural resources could be substituted indefinitely by built capital.[8] Now, if natural resources can be perfectly substituted by capital, when a given resource becomes scarce, it would be enough to reduce its participation in the production process, replacing it with a different combination of labour and capital.

Against this perspective of substitutability, Ecological Economics proposes a funds and flows approach. The idea is that in any production process there are always fund-service resources, i.e., those elements that do not change (in availability) in the short term, and stock-flow resources, which are the elements flowing through the production process.[9] Ecological economists list natural assets, human resources, and means of production among the fund-service resource elements. Rather problematically, they also name these three elements: natural capital, human capital, and built capital, respectively. Stock-flow resources are basically of two types: input flows of materials and energy and output flows of end products and material and energy waste.

Despite the problems briefly mentioned, the spirit of this critique of the idea of substitutability in Environmental Economics raises a crucial point. Machines, equipment (and built capital in general), may only replace elements of nature *within certain narrow limits.*

This brings us to the fourth and penultimate point: the perspective on technology. While exploring the previous point, it becomes clearer that behind the possibility of substitution lies some expectation regarding the power of technology to repair the damage and compensate for the scarcities generated by the economic process.

This implies some level of technological optimism, which in Environmental Economics reaches its highest. First, because of perfect factor substitutability, assumed to be always possible. Second, because the existence of the technologies necessary for adequate substitution is believed to depend only on accumulated knowledge, human ingenuity, and relative prices.

Here is a quick clarification on relative prices for non-economists. A few paragraphs above, the rationality of the economic agent found in Environmental Economics was described as focused on generating the highest possible profit. There are basically two combined ways of achieving this goal: maximising revenues and minimising costs. The second aspect, minimising costs, is more closely related to the topic. Any combination of factors to be used has a cost structure associated with it. The combination that provides the lowest possible cost is assumed to be the most efficient. Notice that the concept of efficiency is entirely subordinate to the objective of maximum profit. From the point of view of this type of rationality, any useful effects of a new technology will not matter if it is more expensive than other available alternatives. In this case, the technology is not economically viable because it is not cost-effective. This is why relative prices matter, even for this highly optimistic view. New technologies only reach economic viability when they become relatively cheaper than the already available technologies (or resources).

The energy sector is a perfect example of how this plays out. It has long been common knowledge that the environmental impact of using fossil fuels is immense. We have long been aware of energy alternatives with a lower, or even considerably lower, impact. These alternatives, however, reach economic viability only to the extent that fossil fuels become more expensive on the one hand and alternative sources become cheaper on the other.

Finally, returning to the explanation thread, this kind of technological optimism present in Environmental Economics poses the issue of sustainability in a very peculiar way. If the elements of the production process are fully substitutable, even in the event of the exhaustion or destruction of any natural elements, the question of sustainability *comes down to finding technological solutions to sustain the growth of the production of goods and services.*

The fundamental point of the ecological economists' critique of this type of optimism is that it fails to recognise the qualitative differences between fund-service and stock-flow resources. In other words, technological optimism loses sight of the fact that productive transformations of the economic process *always* depend on low-entropy resources and *always* generate high-entropy residues. In other words, not even the most miraculous technologies, in the most favourable economic contexts, would be able to fully perform what Environmental Economics expects of them. This brings the idea of limits to the centre of the debate.

Thus, we arrive at the fifth and final point: the perspective on economic growth. Considering everything discussed so far, it can immediately be stated that the possibility of perpetual economic growth is a non-issue in Environmental Economics. In other words, according to its theoretical framework, growth will always be possible, and our technological prowess will clean up our tracks. Herman Daly, a leading figure in Ecological Economics, calls this perspective *economic imperialism*.[10]

In Ecological Economics, there are at least three fundamental approaches in this regard, but they have something in common that must be analysed first. This tradition admits that the material reproduction of humanity depends (roughly speaking) on terrestrial stocks of matter and energy, which are limited, and on the flow of energy from the Sun. This flow is, from a practical point of view, unlimited in total quantity. However, it is limited at specific moments and places, just as the rate at which this energy enters the system is limited. Another aspect of the material reproduction of humanity is the output of the open system, the economy. The waste from our activities tends to deteriorate the environment and the conditions for the material reproduction of human life and other species.

In short, the synthesis of the general process of production and reproduction takes us back to everything discussed so far. It requires material and energy resources on one side and generates waste (also material and energy) on the other. Naturally, when talking about growth, we are talking about an expanding process of production and reproduction. In this respect, economic growth, therefore, demands increasing amounts of materials and energy on one side and releases increasing volumes of waste on the other.

Starting from this fundamental conclusion, we may now look at the three perspectives on growth, all of which are part of the theoretical ethos of Ecological Economics. The first of these is usually referred to as the Spaceship Economy. In it, the spaceship is a metaphor for the Earth as a closed system, representing the objective limits to the expansion of human activity. From this unfolds a defence that the available natural resources should only be used to the extent necessary to maintain the existing stock of capital. In other words, it proposes non-growth. The overall aim is to reduce the socio-environmental metabolic flow. Technology would fulfil this aspiration, supposedly by providing more efficient consumption of natural resources. The rationale is that the higher the efficiency level, the less matter and energy enter the system and, consequently, the less waste leaves it.[11]

The second view is directly connected to Georgescu-Roegen's contributions. The basic idea is that the assumption of perpetual economic growth violates the laws of thermodynamics, especially the second law relating to entropy. Another statement that the thinker derives from the second law is that the amount of matter and energy incorporated into final goods and services is always less than that used during production. In this sense, by producing, we always exchange low-entropy content for high-entropy content.[12] Once again, the problems of how this formulation extrapolates the meaning of the second law must be overlooked to visualise better what those who adhere to these ideas are trying to say.

How much low entropy is lost in the production process would depend on the technological level. Notice that this is a kind of technological optimism but much less pronounced than the optimism found in Environmental Economics. The higher the general level of efficiency in the production process, the lower the proportion of matter and energy wasted as high-entropy elements. There are two positions within this second view. The one that admits that the power of efficiency to avoid this waste generation has narrow limits, drawing from this recognition a defence of *degrowth*. And the position that, while admitting the inescapable nature of the second law, argues that our technological prowess could dematerialise production. What the term dematerialise means, in the end, is the following: we could continue to grow but with a decreasing material impact on the planet. The specific critique of these two positions will be adequately developed in the second part of this book.

Herman Daly, a disciple of Georgescu-Roegen and one of the most prestigious thinkers in Ecological Economics, presents the third and final perspective on growth. Daly is closer, however, to the idea of the Spaceship Economy. He defends a notion of stationarity borrowed from biology.

In the stationary state proposed by the author, natural resources are used not to increase production but to improve the quality of the goods available. The metaphor always used to illustrate this concept is that of a full library. Because it is full, it can no longer expand its collection. However, the collection can still be improved. It would be enough for each advance in incorporating new and better books to be offset by a retreat in that part of the collection that has become old, useless, or outdated. Thus, Daly elaborates a defence of non-growth.[13] However, the arrangement illustrated by the metaphor of the full library is not enough to circumvent the limits imposed by the Ecological Economics' understanding of the socio-environmental metabolic flow and the permanent increase in entropy. Once again, what underpins this conception is faith in technological progress, basically the increase in efficiency levels.

Through the five topics covered, it is possible to draw a reasonably complete picture of the two hegemonic traditions of environmental thinking in

Economics. Of course, being a perspective that is much more compatible with the reproductive logic of capital, Environmental Economics has greater predominance. However, as we will soon discuss, despite its critical origins, Ecological Economics has accumulated many significant similarities with Environmental Economics along its development.

Now, a summary of this chapter can offer points of support for the rest of the book. Traditional economic thinking on environmental issues develops along well-defined lines from what was covered. The four main axes of reflection are (i) the logic of the market, (ii) the sphere of production, (iii) the sphere of consumption, and (iv) the technological dynamics.

The first axis is, among the four, the most self-centred and, to some extent, independent of the others. It includes discussions related to the diagnosis of environmental problems as well as those related to strategies for managing them. In short, it seeks to demonstrate that problems are caused by the absence of market mechanisms or by external interference in such mechanisms.

Suppose this is the origin of the problem. In that case, the political implication naturally becomes the defence of either the creation of these mechanisms where they do not yet exist or the suspension of any external interference. To provide two quick and obvious examples, in the first case, there is the defence of privatisation (of parks, forests, water resources, and more). In the second case, there is the defence of less State interference, trusting that a free market operation would be enough to generate the best possible results, even from an ecological point of view.

While being self-centred, this first axis crosses all the other three because none of the discussions covered in this chapter foresee the possibility of a society beyond capitalism. In other words, they all presuppose capitalism and its "natural," superior, and insurmountable character. Therefore, their entire theoretical and political horizon is circumscribed to the reproductive parameters of capitalism, precisely the mercantile society par excellence, i.e., that society in which the market is, in fact, the centre of social life. So it should not come as a surprise that even in some of the most critical moments of Ecological Economics, the market (its imperatives and its rationality) creeps into the formulations. A critique of this type of perspective can be found in Chapter 7.

The second axis, relating to the sphere of production, usually reconciles proposals that revolve around degrowth, zero growth, or dematerialisation. For reasons discussed in more detail in Chapter 3, positions advocating degrowth or stationarity are pretty marginal. The defence of dematerialisation predominates, aiming to combine economic growth with a decreasing environmental impact.

As for the third axis, the one on technological dynamics, although technological development is multidimensional, considerations fundamentally revolve around the concept of efficiency. Other forms of technological progress are simply abstracted or neglected. Chapter 4 will present the reasons for this.

Finally, in the fourth axis, the formulations relating to the sphere of consumption are usually directed at the general concept of consumption patterns and the exhortation to adopt consumption patterns that generate less ecological impact. Chapters 5 and 6 discuss reasons for the emphasis on individual and voluntary reformulation of consumption habits.

For now, notice how the emphases on each of these axes are related to each other in such a way as to point in a specific direction. It is possible to propose an "equation" that summarises this articulation: the combination of raising the general level of efficiency and new consumption patterns produces a specific result, which is dematerialisation. This relationship not only summarises a way of understanding the interaction between production, consumption, and technology present in both Environmental and Ecological Economics. Above all, it indicates that both converge on the same profession of faith in dematerialisation.

This is the starting point of the stream of thought presented through the next part of the book. Despite being patently unfounded, this bet on the possibility of dematerialisation acquires social circulation because it is functional. But functional for what? To guarantee, even if only on the level of ideas, the reproduction of a society that cannot, due to its structural incapacity, control (limit) the self-expanding tendency of production.

With absolute and relative ecological limits on the one hand and the unlimited drive for growth of capitalist production on the other, only a concept like dematerialisation may conceal this contradiction.

This helps to explain the wide social circulation of such formulations despite their mystified character. Of course, the reasons why these ideas are mystified are yet to be explained, which will be done in Part II. The last part of this introduction presents the structure of the reasoning developed in the next five chapters.

In Environmental Economics and Ecological Economics, with a few exceptions, economic growth is taken a priori as a necessity, as something natural to social life. Supposedly, this economic growth would not necessarily lead to a material expansion of productive consumption. In order to criticise this idea, the factual basis of this mystification must be demonstrated: economic growth *is a necessity*. However, it is a *historically circumscribed* necessity *specific to capitalist society.*

Technological progress would be the main cause of the *supposed* decoupling between economic growth and material impact. In this case, the factual basis of the mystification is that efficiency gains *do* create the *possibility* of saving resources and, at least in theory, producing more with less. Subsequently, it must be demonstrated that various forms of technology are not efficiency-enhancing and that even the saving possibility provided by greater efficiency is not generally fulfilled.

When it comes to consumption, indeed, it is always possible to make changes to individual patterns. It is always possible to choose to consume less or differently. This is the factual basis of this mystification that believes it is possible for these individual choices to become widespread in a sustainable direction. A critique must then be able to demonstrate that the dynamics of capital demand (and therefore actively produce and reproduce) wasteful consumption patterns.

Finally, the belief that market mechanisms drive society to find the most efficient productive arrangements has its factual basis in the fact that competitive pressures effectively demand from individuals a type of rationality that seeks efficiency. Our critique should be able to demonstrate that this efficiency is economic in nature and does not coincide with any desirable notion of ecological efficiency.

Notes

1 This organisation into five topics is based on Cechin & Veiga, *"O fundamento central da economia ecológica" ["The core foundation of Ecological economics"]* (2010).
2 Smil, *Energy: a beginner's guide* (2006).
3 Georgescu-Roegen, *O decrescimento: entropia, ecologia, economia ["Degrowth: entropy, ecology, economy"]* (2008); Daly & Farley, *Ecological economics* (2010).
4 Georgescu-Roegen, *O decrescimento: entropia, ecologia, economia ["Degrowth: entropy, ecology, economy']* (2008).
5 Cechin & Veiga, *"O fundamento central da economia ecológica"["The core foundation of Ecological economics"]* (2010).
6 Smil, *Energy: a beginner's guide* (2006).
7 Ibid.
8 What economists generally call capital is something quite different from what Marx calls capital. The word is the same, but the content is radically different.
9 Daly & Farley, *Ecological economics* (2010).
10 Ibid.
11 Ibid.
12 Georgescu-Roegen, *O decrescimento: entropia, ecologia, economia ["Degrowth: entropy, ecology, economy"]* (2008).
13 Daly & Farley, *Ecological economics* (2010).

References

Cechin, A., & Veiga, J. (2010). O fundamento central da economia ecológica. In P. May, *Economia do meio ambiente: teoria e prática.* São Paulo: Elsevier.
Daly, H., & Farley, J. (2010). *Ecological economics: principles and applications.* Washington: Island Press.
Georgescu-Roegen, N. (2008). *O decrescimento: entropia, ecologia, economia.* Lisbon: Instituto Piaget.
Smil, V. (2006). *Energy: a beginner's guide.* Oxford: One World Publications.

Part I

An ecological critique of capitalism

2 Metabolism and the metabolic rift

The tradition of ecological thinking within Marxism has a long history. There have been at least three stages of what has come to be called ecosocialism. In the first stage, efforts were made to incorporate existing ecological theories into Marxist thought. In the second, there was a predominant effort to find original contributions to environmental issues in the seminal works of Marxism (particularly those of Marx himself). Finally, the third stage is characterised by an attempt to incorporate into the debate the experiences of social movements at the forefront of ecological struggles.[1] As a result, we have a vast and rich literature in which we find very fruitful reflections, not only for the critique of capitalism but also for the prefiguration of a post-capitalist society in effective coevolutionary balance with nature.

One of the contributions most widely circulated among ecosocialists is *Marx's Ecology*, a book by John Bellamy Foster. In this book, Foster practically inaugurates a tradition, best described as an effort to inventory the whole of Marx's work in search of the moments in which the author displays an ecological sensibility ahead of his time, providing insights that are still relevant today. The most striking discovery presented by Foster is the Marxian treatment of the relationship between human societies and the earth, easily expanded to a broad conception of the relationship between humanity and nature.[2]

As argued by Foster, Marx rejects the instrumentalist conception that takes the natural world as a mere static stage for human prowess or as a kind of warehouse with food and objects to satisfy our needs. Instead, he understands that humanity and nature interact analogously to a metabolic process. More importantly than this very appropriate analogy, Marx demonstrates that the dynamics of capitalist society tend to cause a metabolic rift, i.e., it tends to disrupt the metabolic processes that sustain societies and, ultimately, life.

Clearly, Foster's philological discovery significantly impacts the ecological debate, especially on the Marxist side. It makes it possible to bolster that, already in Marx, there are elements to affirm the unsustainable character of the metabolism governed by capital. This might explain a certain frisson around

DOI: 10.4324/9781003512516-3

this Marxian insight, to the point where an actual school of thought has been organised around the discussion of metabolic rift.[3]

However, this kind of infatuation with the notion of metabolic rift has two problematic effects. On the one hand, there has been a progressive *deflating of its critical power* caused by the proliferation of a careless, diffuse, unspecific, arbitrary, and not very rigorous use of the term. On the other hand, the almost exclusive emphasis on metabolic discussion loses sight of other very rich aspects of Marx's work for ecological issues.

This first part of the chapter attempts to underline the precise meaning of Marx's discovery, which Foster retrieved. It particularly emphasises that it already carries a substantive critique of technology (when subject to the logic of capital).

Before discussing metabolic rift, in Chapter 13 of Book I of *Capital*, Marx uses the term metabolism in at least two different circumstances: social metabolism (of commodities, labour, between nations, etc.) and metabolism between humanity and the earth. In both cases, the meaning sought in the analogy is very clear: Marx is emphasising the *processual* character of the objects analysed, their dynamic nature, permeated by flows, interactions, geneses, developments, degradations, depositions, permanences, and changes.

This metabolism through which human societies reproduce themselves, materially and socially, is disrupted by a growing geographical distance between town and country. In general, cities concentrate greater volume and variety of economic activities, better and more extensive infrastructure, health services, education, and more. However, a crucial aspect of this wide range of activities is that almost all of them are non-agricultural, meaning that urban demand for agricultural products needs to be met by a production outside its geographical boundaries. The scale at which agricultural products are shipped from the country to the town tends to keep pace with the urbanisation of the population, since such processes involve an ever-increasing share of the population moving to urban centres and working on non-agricultural activities. The growing geographical distance between typically agricultural activities and typically urban activities creates bottlenecks and interruptions in the return of nutrients and other substances to their place of origin.

This unilateral movement of nutrients, chemical elements, and other substances incorporated into products from the country towards the town has two direct effects. First, it leads to soil impoverishment and loss of fertility due to the soil's nutrient cycle breakdown. The difficulties in replenishing nutrients consumed in production are because, with cities growing, most agricultural product consumption takes place in locations far from their production centres. The international trade in agricultural commodities increases this distance to its maximum.

Second, the non-recycling of animal manure and human waste results in its polluting concentration in urban centres. A genuinely sustainable urban-agricultural system would require the organic recycling of the nutrients contained in sewage and their return to the place of origin, which, as already stressed, is made difficult by the growing geographical distance between towns and the country as towns grow and, therefore, their demand for agricultural products.

These two effects are rarely treated as integrated processes resulting from the metabolic rift. There is, however, an even more severe gap to consider. In the often uncareful allusion to the rift, one almost always loses sight of the explicit critique of technology found in it.

In the face of the tendency for the soil to become impoverished, there is, as is well-known, the possibility of using natural or synthetic fertilisers to compensate for the unilateral movement of nutrients. As early as the mid-19th century, Marx highlighted the following development of agriculture increasingly dependent on new sources of fertilisers:

> Capitalist production collects the population together in great centres, and causes the urban population to achieve an ever-growing preponderance. This has two results. On the one hand it concentrates the historical motive power of society; on the other hand, it disturbs the metabolic interaction between man and the earth, i.e. it prevents the return to the soil of its constituent elements consumed by man in the form of food and clothing ; hence it hinders the operation of the eternal natural condition for the lasting fertility of the soil.[...] But by destroying the circumstances surrounding that metabolism, which originated in a merely natural and spontaneous fashion, it compels its systematic restoration [...]. In modern agriculture, as in urban industry, the increase in the productivity and the mobility of labour is purchased at the cost of laying waste and debilitating labour-power itself. Moreover, all progress in capitalist agriculture is a progress in the art, not only of robbing the worker, but of robbing the soil; all progress in increasing the fertility of the soil for a given time is a progress towards ruining the more longlasting sources of that fertility.[4]

In other words, when the capacity to restore soil fertility grows, it creates not only the possibility of recovering it but also of expanding productivity (volume of production per cultivated area) or even using more aggressive agricultural techniques, the best-known example being monoculture without crop rotation. However, the economic exploitation of these possibilities increases the rate at which the soil is depleted, reproducing the need to restore fertility on a larger scale. This is why Marx concludes, in a passage following the one quoted above, "Capitalist production, therefore, only develops the techniques and the degree of combination of the social process of production by simultaneously undermining the original sources of all wealth – the soil and the worker."[5]

The development of fertility restoration technologies *thus becomes a destructive force*, resulting not in rational agriculture at lower costs but in the expansion of the possibilities for profitable exploitation of the soil. In this case, technical progress makes it possible to guarantee the immediate material conditions of production but tends to reproduce their replacement requirements at an expanding scale. Furthermore, by enabling the expansion (even if only momentarily) of soil productivity, it enables the concentration of ever larger contingents of the population in urban centres, involved in non-agricultural activities, throwing the metabolic rift into a feedback loop.

This reasoning alone provides enough elements to affirm that it is possible to broaden the critical scope of discussions around the metabolic rift. As developed in the next five chapters, the categorical apparatus presented by Marx throughout *Capital* is precisely what makes it possible to base a more comprehensive and radical ecological critique of capitalism. Before that, we can conclude our discussion of metabolism with a crucial recent development in philological research.

For a long time, even among Marxists, there was a prevailing view that Marx's theoretical legacy had little or nothing to offer to reflections on ecology. Until today, influential theorists, such as Löwy, insist that there are anti-ecological elements in Marx's work.[6] Despite the relative persistence of this type of criticism, at least since the 1990s, important works by John Bellamy Foster and Paul Burkett have shown beyond any doubt that Marx often displays a sensitivity to ecological issues which is ahead of his time and that his critique of capitalism has an abundance of very fruitful theoretical elements for the elaboration of an ecological critique of capitalist society.[7]

Foster, in particular, develops exhaustive philological work and, among many other fascinating insights, recovers, as mentioned earlier, notions of metabolism and metabolic rift employed by Marx in *Capital* and the *Grundrisse*. This discovery has undoubtedly been the gateway for generations of ecosocialists since *Marx's Ecology* was published.

From this work onwards, Foster claimed that the metabolic question was not simply a "lost" insight among many others but a central component of Marxian reflection on capitalism. This claim may strike an experienced reader as an exaggeration, specially since Foster does not present enough elements to back up such a strong statement. On the other hand, Kohei Saito's award-winning 2017 book, *Karl Marx's Ecosocialism*, comes a lot closer to fulfilling Foster's ambition. Saito rigorously demonstrates, with textual and biographical support, that Marx's path of investigation and theoretical elaboration was, towards the end of his life, guided by a clarity that the ecological destructiveness of the capitalist system should also be part of his critique of this society.[8]

Fetishism, alienation, and estrangement

In his introduction and first chapter, Saito already rejects the old division between a young, philosophical Marx and an old, more scientific Marx. This periodisation, as he rightly points out, has wrongly placed entire traditions of Marxism either on the side of an overvaluation of the discussion of fetishism, alienation, and estrangement, or on the side of an overvaluation of economic discussions and the emphasis on exploitation. The author correctly realises that Marx's understanding of fetishism, alienation, and estrangement marks his entire intellectual trajectory and serves as the theoretical foundation of his critique up to his mature works.[9]

He opens this discussion by recovering a theme that Foster had already insisted on: the unity between humanity and nature and the rift of this unity. However, while Foster demonstrated in a strictly philological way that Marx was very attentive to this issue, Saito manages to integrate it more organically into the general framework of Marx's understanding of the fetishistic character of capitalist society.

The commodity fetishism section, which closes the first chapter of *Capital*, is well-known. Almost everyone who has heard of this section has probably also heard of one of its most emblematic statements, which affirms that in a society of commodity producers, people relate to each other through things.[10] Going beyond this summary statement, Saito emphasises at least two other crucial points for ecological thinking.

First, he shows how the logical and historical precondition for a society of commodity producers is precisely the *separation* between humanity and nature. The reasoning is simple enough but will be presented in an even more telegraphic form here. As Marx shows us, the generalisation of the mercantile structure of production requires the generalised existence of the free worker. Free from the bonds of domination that mark relations of servitude or slavery, on the one hand, and free from the means of production, on the other. The long historical process of separation between free workers and the means of production involved, in particular, *a profound process of separation between workers and the land*, a process of converting forms of communal ownership of the land into forms of private property, a process of exodus from the country towards the towns; and a process of alienation precisely from the everyday reproductive metabolic relationship between human beings and the land.

The corollary of this reasoning is not just a picture of the disunion between humanity and nature. It is the striking conclusion that in capitalism, this relationship cannot *be any different*. Separation is part of the structural foundations of this society.

The second point, even more profound, concerns the estranged dynamics of the social metabolism governed by capital. In simpler terms, it concerns the fact that that mercantile structure, which is the objective foundation of fetishism, gives rise to a social dynamic that is not only beyond our control but also subordinates us. See how Saito concludes this section:

> From the alien character of labor activity, which is inevitably caused by the estrangement of the worker's subjective capacity in the production process organized by capital, the alien character of the objective world is also produced because labor can only produce products of its own realization as an alien reality. The producers cannot appropriate the product of labor; under a reified dominion, their own activity only realizes itself as a subjugating alien power. This process of de-realization and impoverishment, together with accumulation of capital, produces a constantly growing alien world beyond human control.[11]

Here we have a decisive understanding, capable of providing solid foundations for demonstrating the ecological infeasibility of capitalism. If Marx is correct and if Saito's reading is correct, we must acknowledge the following: the conclusion that the mercantile character of capitalist society generates an autonomised social dynamic that is beyond our control must be accompanied by the recognition that *there are narrow objective limits to managing the development trajectories of such society.*

From an ecological point of view, the importance of this acknowledgement is evident. The fact that it has become necessary and urgent to regulate our metabolism with nature in a conscious and rational (sustainable) way does not mean that this task is achievable within the framework of a capitalist society. Indeed, Marx enables us to demonstrate that the estranged logic of capital blocks every single path to preserve any condition of habitability on the planet.

Indeed, Saito's book offers us many elements to reinforce this conclusion. They will be further explored below.

Social form, materiality, and metabolism

From the first chapter of *Capital*, Marx shows us that wealth in a capitalist society has a double and contradictory character. He starts from the commodity, the "elementary form" of this wealth, emphasising that it has a useful character that makes it a *use value*. This is a general (trans-historical) aspect of wealth. The commodity form, however, carries a specificity: it is wealth produced to be sold, exchanged. The proportion in which it is exchanged for other commodities is its *exchange value*. However, the fact that commodities are equalised in defined proportions indicates that they share a common content, which makes this equalisation possible. This common content, says Marx, is *value*, objectified human labour. Not just any labour, but the common content of all the different types of human labour, i.e., labour reduced to the mere homogeneous expenditure of labour capacity. Marx calls this *abstract labour* and, further on, states that value (this existence of the commodity as mere objectified abstract labour) has a *purely social* existence.

In this way, the dual character of the commodity can be stated by saying that it is both a sensuous object (i.e., a use value, a useful object accessible to the senses) and a supersensuous object (i.e., a value whose existence is purely social).

Saito points out that since this suprasensuous character of value is a specific feature of capitalist society, many have interpreted Marx's theory of value as detached from materiality, a theory that somehow dispensed with the material world.[12] It is even said that capital is *indifferent* to use value. In other words, that capital is indifferent to materiality, that the material form it takes does not matter (throughout the production process or as the final commodity) as long as surplus value is produced. In more precarious versions of this idea, it is sometimes said that the material form of capital does not matter, full stop.

This understanding could not be more mistaken. If capital were really indifferent to its material form, then it would be within our power to seek a materiality compatible with the ecological requirements of sustainability while still within the framework of capitalism.

However, capital is anything but indifferent to its material form. The exact opposite is true. Saito demonstrates that the abstract logic of capital systematically imprints the material world with the forms most suited to its reproduction. Therefore, the actual indifference refers to that fact that the material world produced and developed by the abstract logic of capital is unsustainable, destructive, or ecocidal. It does not matter to capital that its physical existence progressively destroys the natural conditions on which societies are built as long as the greatest possible mass of surplus value is created.

It is important to emphasise that, at this point, Saito is rigorously reading Marx himself. To provide a more obvious example, the entire section of Book I dedicated to relative surplus value is a detailed exercise in demonstrating how capital needs to continually revolutionise its technical base of production in order to overcome the barriers imposed on the extraction of surplus value. The abstract logic of capital demands the elevation of the productive forces, which Marx maps out in the Co-operation, Manufacture, and Modern Industry chapters.[13] It is the uninterrupted rise of these productive forces that needs to be fuelled by exponentially increasing volumes of resources, that dumps exponentially increasing volumes of waste, taking the interaction between humanity and nature in an unequivocal direction, that of persistently increasing disruption and destruction.

In short, Marx's theory of value is not only inseparable from the material world. *It allows us to support the thesis that life governed by value cannot be anything other than destructive*. Saito directly supports this understanding of the relationship between the abstract and the material in Marx's work and shows us that Marx himself was vividly aware of it.

The centrality of the concept of metabolism and the plunge into the natural sciences

When Foster rediscovered Marx's use of the concept of metabolism in *Capital* and the *Grundrisse*, he opened up a fruitful avenue of Marxist ecological thinking. As previously mentioned, however, his demonstration that the metabolic question occupies a prominent position in Marxian thought as a whole is not convincing. Saito collated an even larger volume of texts than Foster had already researched. This way, he managed to come closer to completing this demonstration in a definitive way.

Having presented in detail the importance of the autonomised dynamic of capital and how this dynamic imprint (in a determined direction) transformations on the material world, Saito makes us realise that the metabolic rift, which first appears in Chapter 13 of Book I, is not a standalone side comment. On the contrary, it is a necessary development in the process in which capital is transforming the world to make it increasingly suited to its logic. An increasingly degraded and hostile world, true. Nevertheless, a world better suited to the logic of capital.

When Marx concludes that the separation between town and country, the disruption of the soil's nutrient cycle, and the increasing degradation of soil fertility are *unavoidable* consequences of this blind dynamic, it becomes clear to him that there is a need to understand more deeply the laws that govern the natural processes that make up the ineliminable material basis of the metabolism between society and nature.

Saito maintains this interpretation with extensive textual support from letters and notebooks. But above all, the author reinterprets the plunge into the natural sciences to which Marx dedicated himself from a certain point in his life. This has long been a reasonably well-known fact, but it has always been attributed to the totalising ambition of Marx's scientific endeavour. Saito goes one step further and manages to trace the connection between this turn towards the natural sciences and the specific question about metabolism that Marx came up against throughout his theoretical trajectory.

In other words, Saito convincingly argues that Marx realised the ecological implications of his critique of political economy and actively sought to complete it by becoming acquainted with the controversies on the frontiers of soil science, geology, mineralogy, and more.

Indeed, this endeavour did not result in an autonomous reflection by Marx on ecological issues. Even so, it is clear that the theoretical legacy he left behind is not only a rich source for the ecological critique of capitalism. It is clear that he himself took the first steps in this direction. Hence, the title of Saito's work, *Karl Marx's Ecosocialism: Capital, Nature, and the Unfinished Critique of Political Economy*, is perfectly appropriate. On the one hand, the author's critique of political economy did not receive the necessary in-depth analysis of ecological issues. On the other hand, Marx himself had already realised the need for this in-depth study. In pursuing this line, he pioneered a whole tradition of ecological thought influenced by him, which would only flourish in the latter part of the 20th century.

Notes

1 This division of ecosocialist thinking into phases will be discussed in more detail in Chapter 8.
2 Foster, *Marx's ecology* (2000).
3 White, Gareau, and Rudy refer to the authors of this tradition as *rift scholars*. See White, Gareau, & Rudy, "Ecosocialisms, past, present and future" (2017).
4 Marx, *Capital*, vol. 1 (1982), 637.
5 Ibid., 638.
6 Löwy's critique will be addressed in Chapter 10.
7 See Foster, *Marx's ecology* (2000); and Burkett, *Marx and nature* (1999).
8 Saito, *Karl Marx's ecosocialism* (2017).
9 In a recently published book, there is an exquisite chapter by Bianca Bonente and Hugo Corrêa that revisits this debate and undoes this sterile dichotomy. See Bonente & Corrêa, "Entre o fetichismo e a exploração" ["Between fetishism and exploitation"] (2021).
10 Marx, *Capital*, vol. 1 (1982), 163–177.
11 Saito, *Karl Marx's ecosocialism* (2017), 48.
12 Ibid.
13 Marx, *Capital*, vol. 1 (1982).

References

Bonente, B., & Corrêa, H. (2021). Entre o fetichismo e a exploração. In J. L. Medeiros, & E. Sá Barreto, *Para que leiam O Capital: interpretações sobre o Livro I* (pp. 133–166). São Paulo: Usina Editorial.

Burkett, P. (1999). *Marx and nature: a red and green perspective*. New York: St. Martin's Press.

Foster, J. B. (2000). *Marx's ecology*. New York: Monthly Review Press.

Marx, K. (1982). *Capital* (Vol. 1). London: Penguin Books.

Saito, K. (2017). *Karl Marx's ecosocialism: capital, nature, and the unfinished critique of political economy*. New York: Monthly Review Press.

White, D. F., Gareau, B. J., & Rudy, A. P. (2017). Ecosocialisms, past, present and future: from the metabolic rift to a reconstructive, dynamic and hybrid ecosocialism. *Capitalism, Nature, Socialism, 28*(2), 22–40.

3 The insatiable drive for expansion of capital

This chapter attempts to expose how the dynamics of production governed by capital are necessarily driven to growth. Three distinct moments make up this reasoning. It starts with an analysis of value as a category of exchange. At this point, the aim is to emphasise how we can identify a drive to expand production just by admitting the simple circulation of commodities. The level of abstraction will still be relatively high, which means that some very important categories of the system will still be absent from the analysis (capital itself, for example). The form of circulation that characterises this stage of the analysis is C-M-C, i.e., a commodity is exchanged for money that is exchanged for a commodity. The second part of the presentation deals with value as capital. At this point, the focus shifts to the distinction between value as the mere quantitative objective existence of the commodity and the value that begins to circulate as capital. Subsequently, it is highlighted that not only does its form of circulation change from C-M-C to M-C-M'; i.e., money becomes a commodity that becomes more money. The content of this circulation is also radically different. Finally, we will discuss capital in its movement of expansion.

This last point already illustrates the first connection between the movement of capital in its value dimension and the movement in its use value dimension. That will, therefore, give an inkling of the connection between the abstract dynamics of capital and the growing material impacts of capitalist production. But this is only the first step. It is not enough to definitively rule out the idea of dematerialisation. That will require a specific discussion of technology, which will be in the next chapter.

If all this theoretical elaboration is built from a Marxian perspective, it must start with a return to the analysis based on value. For those who have not heard (or do not remember) how Marx demonstrates the origin of value in labour, its reasoning will be reproduced here in a highly telegraphic way.

Marx immediately emphasises that, when exchanging, individuals exchange qualitatively different commodities. This is obvious because exchanging qualitatively equal commodities would make no sense. At the same time, despite this

DOI: 10.4324/9781003512516-4

difference, something must be equalised. This is another prerequisite for the exchange to make sense. Thus, in exchange, a certain quantity of an object is equalised with another quantity of another object. Disregarding all the qualitative specificities of both commodities, the only feature they have in common is that they are both products of human labour.[1]

However, even the labour that produces commodities is qualitatively different. The labour that produces fabric is different from the labour that produces wine, which is different from the labour that produces wooden furniture, etc. Marx calls this qualitative dimension of labour *concrete labour*. It gives rise to the use value of commodities. However, these different types of work share something in common: they expend human labour capacity for a specific period. This common feature, shared by the most varied activities, is what Marx calls *abstract labour*. It is the abstract labour objectified in commodities that subsists in them as value.

In this passage, Marx shows that the labour that produced in order to exchange has a dual character: concrete labour and abstract labour. He also shows that the product of this labour, the commodity, has a dual character: *use value*, associated with the concrete character of labour, and *value*, associated with the abstract character of labour.

A word of warning to avoid any misunderstanding. There is no concrete labour on the one hand or abstract labour on the other. There is commodity-producing labour that is, *at the same time*, concrete and abstract. Similarly, there is no use value on the one hand and value on the other. There is the commodity, which is a dialectical unity of value and use value.

Why is it a dialectical unity? Because it is, potentially, both things at the same time. Nevertheless, the realisation of either of these dimensions requires negating the other. So, when consuming the commodity, one realises it as use value but negates it as value. Likewise, when selling the commodity, one realises it as value but negates it as use value.

Another important point is that Marx is very clear in stating that the use of value is the condition for the existence of value. Even though use value does not determine a commodity as value, it must be the first to be the latter. A commodity whose use value is destroyed for any reason ceases to be a material support for value.

Hopefully, that will be enough to follow the development of the argument easily. Our stream of thought begins with this last point.

Does anyone doubt that, under capitalism, the overwhelming majority of social labour has exchange as its goal, i.e., for the market and not for immediate personal consumption? Anyhow, if the product of labour is systematically directed towards the market, it acquires that dual existence: value and use value. In such circumstances, the product of labour is a commodity.

It is essential to realise that this feature is not only distinctive but *exclusive to capitalism*. Only in this society do the products of labour take the form of commodities in a universal, generalised way. So we have a society in which producers work and produce with the aim of taking their end product to the market. Commodity production is always production with the purpose of selling.

Here, the product of labour is always destined to change hands. This means that it is always a non-use value *for its producer*. In other words, the useful properties of this product do not directly satisfy any of its producer's needs. However, because the commodity needs to be sold, it must be able to satisfy needs other than those of the producer himself; it *must be* a use value for potential buyers.

Therefore, the producers privately produce things that are non-use values to them, take these products to the market, and only then do they obtain (or not) the means to satisfy their own needs. From this, we derive two immediate conclusions. First, each producer confirms their work as an integral part of social labour only by ratifying the necessity and usefulness of their product through the realisation of exchange. That is, producers produce in isolation, and the totality of private/isolated labour is social labour only *potentially*. Private labour only becomes part of social labour if it successfully participates in the market. Therefore, and this is the second conclusion, it matters to the possessor of the commodity to what extent it can be exchanged for other products. This means that the product of one's labour meets one's needs only to the extent that it allows one to claim a given share of the total social wealth.

And how is the magnitude of this share determined? Access to social wealth is, as demonstrated, mediated by exchange, and the proportion in which commodities are exchanged is determined by value. To be more precise, the magnitude of value is determined by the socially necessary labour time.

To close this first part, it is worth summarising the outcomes obtained. First, in exchange, products with different use values are made equivalent in specific proportions. That is possible because of their quality of being objectified human labour, i.e., because they are values. Second, it is value, therefore, as the category of exchange, that establishes the relationship between the labour privately performed by each producer. Third, it is the market that necessarily mediates between individual labour and the social wealth produced.

If individuals equalise the values they possess in the act of exchange and if exchange is the dominant interposition between the individual and the objects needed to meet their needs, then the more value one possesses, the greater the volume and variety of use values one has access to.

As the direct exchange of commodities (barter) is incompatible with the generalised production of commodities, in the course of the historical process of generalising commodity production, there is a parallel development of the forms of expressing value. In other words, the more frequent and widespread the exchange is, the more efficient the way of equalising commodities as values must be. The culmination of this development of the forms of expressing value is money.[2]

One of the fundamental functions of money is to express the value of all commodities. At the same time, all commodities express their value in money. Money, then, is the standard value form of commodities. In other words, it is the representative of value. Money expresses its power of attraction over social wealth, being convertible into any commodity as long as it is presented in the right proportion.

However, individuals may only withdraw money from circulation to the extent that they provide equivalent value in commodities. Therefore, the more one produces, the more one can sell and, consequently, the greater one's access to material wealth. As producers do not produce for themselves but to sell, their production is not limited or determined by their personal needs or those of their dependents. In other words, the goal of their production is not directly meeting specific needs. Instead, the aim is to ensure the means to indirectly fulfil any needs that may arise. In principle, there is no pre-established limit to the amount of money that can be obtained from circulation (momentarily leaving aside the possibility that the value produced is not realised, i.e., that the commodity is not sold).

In any case, even at this very high level of abstraction, there are already enough elements to perceive the drive to expand production originating from this mercantile structure of wealth production. This impulse stems not only from the desire to get rich but also from the need to avoid relative impoverishment. If this is a society in which social wealth tends to grow, immobile, stationary personal wealth represents a magnitude that, in proportional terms, is ever smaller in relation to the growing totality of social wealth.

This is why the C-M-C circuit already carries the seeds of the need to circulate value as capital, which overcomes the limit of immobile wealth. In this society, the *appropriate form of wealth conservation* is, therefore, one that *also drives accumulated wealth towards growth.* This is precisely the subject of the following section: value circulating as capital.

***

The previous section demonstrated that (i) the origin of value is labour; (ii) value mediates exchange; (iii) exchange mediates private labour; (iv) the market mediates private labour and social wealth; (v) what is given in money is taken from the market, but only what is given in commodities is obtained in money; (vi) thus, the greater the production, the greater the amount of money, the greater the access to socially produced wealth. Therefore, the commodity-producing society has an immanent drive to expand production. This whole logic can only be unfolded by considering value as a category of exchange.

By adding new determinations, one may observe value circulating as capital. The first point to emphasise is the change of form. Before, the value circulated in the C-M-C form. However, the typical circulation of value as capital is different, M-C-M'.[3] What circulates in both cases is value. Therefore, this seems to be a mere change of form at first glance. As if, instead of observing the sequence of transactions through a section that starts with the commodity, that exact sequence of transactions was observed through a section that starts with money.

That would be a mistake. This change in form brings with it significant changes in content. In other words, when the nature of the process changes, the characteristics of the movement change. In C-M-C, the ultimate goal of the movement is a specific use value to fulfil a need. Even if considering the drive to expand production presented in the previous section, the final moment of this circuit is always the withdrawal of a commodity from circulation and its realisation as a use value through consumption. In this kind of movement, money constantly changes hands, taking the place of a commodity withdrawn from circulation. This circuit, for each producer, is renewed with the renewal of the need that motivates it. Its repetition is determined by a purpose outside the operation, the use value. Finally, the beginning and end of the process differ *qualitatively*. The commodity that closes the circuit differs from the commodity that starts it.

In M-C-M', the starting point and the "final" goal of the movement change. Exchange value becomes the objective that drives the whole operation. Money starts the circuit and, more importantly, returns to its original holder at the end of the movement. Furthermore, if everything works as intended, the money returns increase in value. If this money does not return (or does not return increased in value), the operation has not been successful. The beginning and end of the process differ only *quantitatively*. As the objective is precisely quantitative expansion, this operation is perpetually repeated. Whenever this repetition is interrupted, the movement characterised by C-M-C takes place again, and that value ceases to be capital. In other words, *value only functions as capital when it circulates, producing its expansion.*

It has been said that the producer, now an owner of capital and therefore a capitalist, advances money and expects to receive it increased at the end of the process. How does this happen? A surplus must be added to the original quantity for this to happen. This surplus is what Marx calls surplus *value*. The return of the money with added surplus value is determined by how it is spent. A capitalist does not accumulate wealth by keeping money out of circulation. A capitalist accumulates wealth by continually launching money into circulation. Nevertheless, he does so in a particular way: this money is spent *as capital.*

And how does one convert value into capital? The first step is to convert part of the value into means of production and part into labour-power. The second is to use these contracted means in a production process under certain conditions.[4] The third is to launch the product (commodities) into circulation.

It is worth examining the capitalist's first trip to the market. Means of production and labour-power exist there as commodities. The capitalist buys means of production from the producer and pays the corresponding value. The capitalist buys labour-power and pays the value corresponding to its original holder (the worker). However, be careful: the capitalist is *not* spending value to buy labour, which is the source of value, thus obtaining a greater mass of value. The capitalist is buying *labour-power*. This is very important. Labour and labour-power are very different things, and Marx, unlike Smith and Ricardo, understands this difference with great rigour. Further along our discussion, it will become more evident that paying for labour-power instead of paying for labour is decisive for the demonstration of surplus value.

Labour-power is a unique commodity, but like any other, it is also a unit of use value and value. And what is the value of labour-power? As with any commodity, it is the labour time that is socially necessary for its reproduction. This is the labour time needed to maintain a worker in the average living conditions determined by social and historical context. It is the sum of the value of all the means necessary for a worker's maintenance and reproduction.

And what is the use value of labour-power? How fortunate for the capitalist: it is labour. If labour-power is the living capacity to work that exists in each person, the need that labour-power satisfies is precisely the exercise of labour. In other words, the same activity that realises labour-power as a use value (labour itself) is the one that creates value.

So this is a unique commodity that, on the one hand, is objectified labour (value) and, on the other hand, is the capacity to objectify labour, i.e., to create value. Thus, the demonstration of surplus value can be formulated *without* assuming that the exchange between capitalist and worker is unequal, in other words, without having to assume that the worker is necessarily paid less than the value of his labour-power.

The labour objectified in the labour-power and the labour it can perform are dissociated, having no causal relationship. That is, the value of labour-power and the value it creates are different quantities. If the productive employment of labour-power creates value, when employed up to a point (say, half the day), it creates enough value to equal what was paid for it. For the rest of the time, it will create surplus value, value beyond that which was advanced for it.

With these additional determinations, the discussion of how value is formed evolves. Indeed, labour creates value. However, is the whole value of the commodity created by the living labour directly involved in its production? No. Past labour, objectified in the means of production, also participates in creating this commodity. Therefore, both the workers and the means of production play a part in the formation of the value of a commodity.

This is *not* to say that the means of production *create* value, but that they have their value transferred to the commodity. This transferred value does not change in magnitude; it passes unchanged from the means of production to the commodity. That is why Marx calls the part of capital that exists as a means of production *constant capital*. The part of constant capital consumed in production reappears unchanged in the value of the end product, the commodity. The rest of the value of the commodity is created by living labour. Precisely because it is a newly created value, Marx calls this part a *new value*.

In summary, the value of the commodity is the sum of the value transferred from constant capital and the new value created by labour in activity. The new value, in turn, may be looked at in more detail. We already know that over the course of an entire day, labour creates a certain amount of value. Until a certain point in the day, this value is only enough to replace what the capitalist has advanced to the worker. Since the part of capital that is advanced to the worker needs to be replaced by the creation of new value, Marx calls this part *variable capital*. All new value that exceeds variable capital is *surplus value*. It is precisely this excess of new value created that the capitalist seeks: surplus value will guarantee that the amount of value at the end of the process will be greater than at the beginning.

To facilitate some of the discussions on ecological implications developed below, consider the value composition as the sum of constant capital, variable capital, and surplus value. Alternatively, to use its most straightforward notation, $c+v+s$.

Before exploring the ecological implications, this theoretical review has more to cover. This section mentioned that the characteristic circuit of capital is M-C-M' and that the expansion of money depends on applying value as capital. This involves contracting means of production and labour-power, their employment in a productive process, and the launching of the product into circulation. Furthermore, the value of the commodity launched into the market is made up of a portion that retains the value of the means of production, a portion that returns/restores the value paid for the labour-power, and a portion that exceeds the value advanced.

At this level of abstraction, two fundamental ways of expanding surplus value (something that obviously interests the capitalist) may be discussed. One can extend the working day or reduce the need to pay for labour-power by reducing its value. With regard to this second way, it should be pointed out that it is not a question of paying labour-power below its value but of reducing the value of labour-power itself.[5]

The first way to increase surplus value is simply by extending the time in which workers work to create value. As the time required for workers to reproduce the value advanced by their labour-power remains the same, any time they work beyond this is surplus value creation time; it is surplus labour time. The extension of the working day is, therefore, the extension of the time in which workers create surplus value. The extraction of surplus value by extending the working day is called *absolute surplus value.*

Before moving on to the second way, a little parenthesis: Marx calls the time that workers need to reproduce the value of their labour-power *necessary labour time*. This means that, during this time, workers produce value equivalent to what they received in wages. In other words, they produce, albeit indirectly, their existence. It is essential not to conflate necessary labour time with *socially* necessary labour time, which determines the magnitude of the value of a commodity.

The second way increases surplus value by reducing the necessary labour time by decreasing the value of labour-power. As a result, even if the working day does not increase, the surplus labour time, when the worker produces surplus value, increases. The extraction of surplus value by reducing the necessary labour time is called *relative* surplus value.

Incidentally, the ratio between surplus and necessary labour time gives us the rate of surplus value, which, according to Marx, tells us the exact degree of exploitation. The rate of surplus value can also be found in the ratio between the values created in a given period, i.e., surplus value over the value of labour-power; or, in a simple notation, s / v.

In principle, extracting absolute surplus value is simpler and more straightforward. All the capitalist has to do is extend the time in which means of production and labour-power are consumed in the production process. However, there are absolute, physiological, legal, and sometimes even moral limits to extending the working day.[6] For this reason, capital, which never has its thirst for surplus value quenched, is forced to revolutionise its technical base to continue the process of expanding the creation and appropriation of surplus value by increasing productivity.

The important thing now is to realise something that has already been present in this discussion but which might have gone unnoticed. If the working day remains unaltered, what happens to the mass of value produced in the face of productivity variations? When the general level[7] of productivity increases, but the working day remains the same, the total amount of value suffers no variation.

A general increase in the level of productivity only lowers the value of labour-power, reducing the need to advance variable capital and consequently expanding surplus value.

Why does the value of labour-power decrease when the general level of productivity rises? Doesn't greater productivity mean it is now possible to achieve the same result in less time? Since it is the general level of productivity that has risen, it is the *socially* necessary labour time to produce one unit of any commodity which has decreased. Thus, the value of commodities decreases, including that of those commodities that form a worker's means of subsistence, the sum of which forms the value of their labour-power.

However, if the value of each commodity decreases, how can the total amount of value created not change when the general level of productivity rises? Because *the number of units produced increases*. If, on the one hand, an increase in productivity means that the same commodity can be produced in less time, on the other hand, it also means that more units can be produced in the same amount of time.

Looking at it from another angle, if the general level of productivity increases, the amount of labour socially necessary to produce each commodity decreases, and, therefore, its unit value decreases as well. If production remained at the same level of units produced, the total value created would decrease. As the aim of capitalist production is to expand value, the quantity of units produced must *necessarily* grow to more than compensate for the decrease in the unit value of the commodities. This has a decisive ecological implication.

Once again, it is worth taking a moment to summarise the sequence of reasoning so far. First, when the value circulates as capital, it follows the M-C-M' circuit. Second, in this circuit, what matters is the extent to which value can be expanded, i.e., what matters is surplus value. Third, there are two fundamental ways of expanding surplus value: absolute surplus value (by extending the working day) and relative surplus value (by increasing productivity). Fourth, as the first form has objective limits, the second form is more than just an alternative; it is a necessity. Fifth, the increase in the general level of productivity has the effect of causing the unit value of the commodities produced to fall. Sixth, as a result, in order for surplus value to continue its course of perpetual expansion of capital, the quantity of units produced *has to grow*.

Theorists of Environmental Economics and Ecological Economics have gotten this far, albeit along completely different paths. However, in these (and so many other) traditions of thought, this drive for growth is naturalised. In other words, the need to grow is accepted as somehow a constituent fact of human nature or life in society. This obsession with growth is not recognised as a specific feature of capitalist society.

Even among leftists, even among communists, it is not uncommon to assume that economic growth is something to be pursued. This theoretical review aims to show the real basis of this mystification: *in capitalism*, growth *really* is an unavoidable necessity. It is no coincidence that the forms of consciousness committed to this society need to seek shelter in the idea of dematerialisation[8] since they cannot give up the imperative of growth.

In the last part of this chapter, conclusions on growth will be reinforced by a brief discussion of capital accumulation, i.e., by observing this growth impetus of capital as a process of recurrent repetition.

A brief deepening of the previous analysis, to observe capital in motion, must precede the following discussion. So far, the reasoning was based on the assumption that the number of workers employed was constant. What else may be discovered by reversing the reasoning and keeping everything else constant except the number of workers? If the length of the working day, the productivity, and the rate of surplus value are constant, the mass of surplus value will depend on the number of workers employed.

In turn, the number of workers employed depends on the physical capacity of the means of production to employ workers. In other words, the means of production have defined technical characteristics that make employing workers in well-defined proportions possible. Even when there is some room for manoeuvre, it is always limited. In addition to these physical/technical characteristics of the means of production, their capacity to employ workers will also depend on the magnitude of their whole. All other things being constant, the greater the mass of means of production, the greater the mass of workers employed. To sum up, assuming this, the number of workers employed depends on what Marx calls the *technical composition of* capital, i.e., the ratio between means of production and labour-power.

How is this connected to the previous presentation? Before, the number of workers was constant. However, the process of increasing productivity tends to unemploy the workforce. If one way of defining productivity is "more production with the same amount of labour," another equally valid way is "the same production with a smaller amount of labour." In other words, for each given level of production, the higher the productivity (i.e., the higher the technical composition), the lower the capacity of the means of production to employ labour.

As Marx shows us, a direct implication of this reasoning is that new branches of production that have already emerged with high productivity have less capacity to attract labour. Moreover, when renewing their productive apparatus, the old branches tend to unemploy workers who had previously been employed.

This demonstrates that the processes described in the second part of this chapter aim to expand the mass of surplus value available for appropriation and, with it, capital. Also, that the technical result of increasing productivity is a reduction in the capacity of the means of production to employ living labour, reducing the participation of the value-creating substance itself and, of course, of surplus value. Therefore, capital must grow *also to compensate for this effect*, i.e., to continue to be able to consume labour-power. In the next chapter, the ways this logic works for global capital will be further analysed, even though this impulse is reversed on the level of individual capital, that is, to increasingly do without labour-power.

But what does it mean to say that capital must grow? Considering that value is only capital as long as it operates *as capital*, if, at the end of M-C-M', the capitalist appropriates the surplus value and consumes it entirely in an unproductive way, the original value may continue to operate as capital, but by simple reproduction. In other words, in theory, this capital would reproduce itself indefinitely in a repetition of M-C-M', always returning to the same base.

Simple reproduction, however, does not include the processes previously described. If there is an increase in productivity, the second M-C-M' does not produce the same mass of value as the first. This is because, with increased productivity, the initial M is distributed in a different proportion between the means of production and the labour-power. It employs less living labour and therefore creates less value.

The point here emphasised is that, if this reasoning is correct, part of the surplus value must necessarily accumulate. In other words, it must become capital, expanding the base on which the movement of valorisation reproduces itself. The capital at the beginning of the circuit must become increasingly larger whenever there are productivity gains. Again, this logic concerns the operation of global capital.

The ecological implications are once again evident. The very nature of capital requires not only increasing value but also increasing amounts of production. The very fulfilment of these demands over time requires the capital in operation to grow constantly. In short, it requires capital accumulation. It requires that the entire logic of growth described in this chapter reproduces itself on an expanded scale, in a spiral. The only rigorous conclusion compatible with all this reasoning is that any defence of degrowth that does not presuppose the overcoming of capitalism suffers from an original defect: it fails to recognise that capitalist society cannot be anything other than ever-expanding.

Marxist literature on degrowth does not make such a primary error. Important recent texts assert that the necessary degrowth (both in its qualitative and quantitative aspects) cannot be realised under capitalism. However, something happens when it comes to the effort of prefiguring what this necessary degrowth would look like. When obliged to outline a so-called degrowth communism, Marxists often allow for the intrusion of elements of this prefigured future into the terrain of what is, in the present, politically viable.

Saito, for example, seems to suppose that the path to degrowth communism is set by degrowth already within capitalism, by multiplying production spaces that prioritise use value, replacing (on this micro-scale) production governed by value.[9] Foster, on the other hand, alludes to the importance of planning for the execution of a degrowth trajectory, but conceives it as planning for the "deaccumulation" of capital, which presupposes (even if unwillingly) that such a path would be pursued still under the rule of capital as the central dynamic vector.[10] Even when his reasoning refers to the Soviet NEP (and, therefore, to a post-capitalist transformation experience), it is crucial to keep in mind that this was only possible after two revolutions and a civil war. Löwy manifests the same confidence in the immediate viability of this line of action by stating that, "without nurturing any illusion about a 'clean capitalism', we must try to buy time and impose on the constituted powers some elementary degrowth measures."[11]

Indeed, none of these authors harbours illusions about the possibilities of green capitalism. Furthermore, it is absolutely correct to extract from the current scientific understanding the imperative of an accelerated planned degrowth by the people. The urgency of the task and the clarity with which the correct direction presents itself leads to an inevitable prefiguration of how the future should be. However, describing what degrowth communism should look like in general terms does not provide a map of the transformations we would need to imprint on the present conditions to make it viable.[12]

In other words, degrowth does not begin to be realised by adopting degrowth measures in the present. The ecological critique of capitalist society demonstrates that there is a dense network of insurmountable objective limits to it. If these are the roots of such impossibility, degrowth may only be brought within reach to the extent that these limits are overcome, through a revolutionary rupture with the society of capital.[13]

Notes

1 Marx, *Capital*, vol. 1 (1982), For a commented presentation, see Sá Barreto, *De Smith a Marx: curso introdutório em dez aulas ["From Smith to Marx: an introductory course in ten lessons"]* (2021), Lesson 07.
2 Marx, *Capital*, vol. 1 (1982), For a commented presentation, see Sá Barreto, *De Smith a Marx: curso introdutório em dez aulas ["From Smith to Marx: an introductory course in ten lessons"]* (2021), Lesson 08.
3 Marx, *Capital*, vol. 1 (1982), For a commented presentation, see Sá Barreto, *De Smith a Marx: curso introdutório em dez aulas ["From Smith to Marx: an introductory course in ten lessons"]* (2021), Lesson 09.
4 These conditions will be discussed in the next chapter.
5 Marx, *Capital*, vol. 1 (1982), For a commented presentation, see Sá Barreto, *De Smith a Marx: curso introdutório em dez aulas ["From Smith to Marx: an introductory course in ten lessons"]* (2021), Lesson 10.
6 The increase in work intensity is momentarily ignored in this chapter. It will be dealt with in Chapter 4, on technology.
7 At this point, the variations considered concern the general level of productivity. The case of specific variations will be dealt with in Chapter 4, on technology.
8 Further discussed in Chapter 4, on technology.
9 Saito, *Slow down* (2024).
10 Foster, "Planned degrowth" (2023).
11 Löwy, "Nine theses on ecosocialist degrowth" (2023).
12 Chen, "Degrowth – what's in a name?" (2023).
13 Cf. Chapters 15 and 16.

References

Chen, Y. (2023). Degrowth – what's in a name? Assessing degrowth's political implications. *Monthly Review, 75*(3), 160–170.
Foster, J. B. (2023). Planned degrowth: Ecosocialism and sustainable human development. *Monthly Review, 75*(3), 1–29.
Löwy, M. (2023). Nine theses on ecosocialist degrowth. *Monthly Review, 75*(3), 154–159.
Marx, K. (1982). *Capital* (Vol. 1). London: Penguin Books.
Sá Barreto, E. (2021). *De Smith a Marx: curso introdutório em dez aulas.* São Paulo: Usina Editorial.
Saito, K. (2024). *Slow down: the degrowth manifesto.* New York: Astra House.

4 Technology and the myth of dematerialisation

This chapter begins to address the link between the conclusions of the last two chapters and technological dynamics. Indeed, some remarkable features of technological dynamics have already been mentioned. Some general tendencies that push capitalists to raise productivity levels under their capital have been analysed. The following paragraphs explore this dynamic in greater detail.

The first step is to carefully define what technological development is and its function within the more general process of raising the productive forces. As mentioned before, traditional literature normally interprets technological development as a rise in efficiency levels, i.e., as an increase in the saving capacity of the technology employed.

This is the first problematic point. The idea of technological development as the development of the resource-saving capacity is imprecise, incomplete, and partial. First of all, in a historical sense. In the history of humanity, *countless* examples of technologies have enabled the exact opposite: an increase in resource-consuming capacity. The steam engine, for example, makes it possible to consume more coal, as it multiplies the possible uses for coal. The evolution of its efficiency reinforces this effect, making it possible to use it where less efficient versions are not functional. The invention of the combustion engine boosted oil consumption. The emergence of electricity transmission networks boosted the consumption of all available energy sources. Oil platforms allow drilling for the resource in previously inaccessible places. More efficient platforms make it possible to maximise the extraction in known reserves. Harvesting machines make it possible to farm on a scale that despoils the soil. The development of synthetic fertilisers makes it possible to expand this scale. The jet turbine is more efficient than the turboprop engine, but it means more fuel consumption because it enables the vertiginous growth of air transport. This rapid exercise in illustration offers a series of obvious examples that do not support the idea that technological progress necessarily points in the direction of moderating our ecological impact.

DOI: 10.4324/9781003512516-5

However, two aspects of this discussion are worth emphasising. First, more is needed to conclude that technological development is moving in the opposite direction, accelerating the impact. Other elements must and will be presented to make such a statement. Second, this limited notion of technological development does not provide the exact meaning of the development of *productive forces*.

As warned, the emphasis on increasing productivity in the previous chapter was still incomplete. Capitalists develop productive forces to expand surplus value. But they do not do this exclusively in the way presented so far, i.e., on the basis of increases in productivity.

One way of visualising the broad process of developing productive forces is by considering two approaches. In the first one, there are two possible origins of this development. It can occur through changes in the *means of production*. In this case, the changes are generally associated with some technological development. However, the development of productive forces may also come from changes in the *organisation of the production process*. Indeed, in our time, significant organisational changes usually presuppose technological developments. Nevertheless, that does not change the fact that they can occur without these developments. Chapter 11 of *Capital* (on cooperation) offers an example of how the simple spatial reorganisation of the production process raises the productive forces. Not to mention Adam Smith and his famous analysis of how the division of labour within the production process has an explosive impact on the level of productivity.[1]

In a second possible approach to the tendency to increase productive forces, the analysis may focus on three results of different natures. The increase in productive forces can increase labour productivity, efficiency, or labour intensity. All three effects can come from technological changes or not.

The point to be emphasised is that capitalist society has a vigorous technological dynamic because the need to expand surplus value drives capitalists to raise the productive forces under their command. Naturally, part of this will drive specifically technological developments. But these developments cannot be reduced to ways of saving resources and thereby supposedly moderating our ecological impact. This process is not even about technological change in its strict sense. In other words, when thoroughly considered, the increase in productive forces is far from being just about more efficient technologies, as common sense or even traditional literature would have us believe.

This very preliminary analysis alone, which only identifies different aspects of the process in question, is enough to challenge the notion of dematerialisation. However, one may deepen the discussion by assessing the main determinants underlying the three results listed: productivity, efficiency, and intensity gains. In doing so, a few outcomes may be achieved. Firstly, to explain their structural role in the general dynamics of capital and the mechanisms driving them. Secondly, to understand why productivity, efficiency, and intensity gains are necessary for the adequate reproduction of the social metabolism governed by capital. In other words, to understand that they do not happen either by chance or driven by some anti-ecological collective consciousness. Finally, to understand why they invariably have a material, ecological, and growing impact.

First, it is essential to understand the crucial difference between productivity and efficiency, which has only been superficially commented on until now. Efficiency is always related to the means of production. The level of efficiency provides us with the ratio between inputs and output. The closer this ratio is to 0, the more efficient the process is. This is what Marx calls the effectiveness of the means of production.[2] Productivity, on the other hand, always refers to labour. The level of productivity provides us with the ratio between the living labour performed and the product. Traditional literature usually presents these two terms as synonyms. However, this imprecise way of dealing with the terms conceals essential differences.

As seen in the last chapter, increasing productivity has the necessary result of increasing the number of units produced. If an increase in efficiency does not accompany this increase in productivity, another result is necessary. Suppose the relationship between inputs and output has not changed. In that case, a productivity gain that results in increased production can only be accompanied by an increase in the consumption of these inputs. In other words, if only labour productivity has increased, but the volume of raw and auxiliary materials consumed at each level of production remains the same, the increase in production will lead to an increase in the consumption of raw and auxiliary materials. Assuming, as it seems fitting to do, that these resources continue to be processed in the same way, a third necessary effect is the growth in waste generation.

On the other hand, increased efficiency reduces input consumption for *any given level of output*. Thus, it creates the *possibility of* material savings. But if output is increasing (as previously demonstrated), nothing can be said *a priori* about the absolute level of resource consumption. Necessary additional elements will be presented further in the discussion.

The increase in productive forces is how capital responds to the limitations on longer working hours, but that is merely the dimension of productivity. The increase in the intensity of labour has only been mentioned, but not in depth. In the same way as productivity gains, increasing the intensity of labour also allows the capitalist to expand surplus value, but not by extending the working day or contracting the necessary working time. The exceptionally intense working day compresses a longer working day into a fixed period. In other words, the most intense working day condenses the labour of a longer working day into a given amount of time.

The same exercise performed for productivity gains may be applied here as well. If the level of efficiency remains constant, a technological development that leads to an increase in labour intensity will necessarily increase the material consumption of the production and, subsequently, in the generation of waste. Since the efficiency level is given, the volume of resources consumed for each given amount of output is also given. As the increase in intensity corresponds to the compression of a longer working day into a given working day, the amount of output will grow, and with it, so will the material impact of production.

Similarly to what occurs when technological developments increase productivity, technological developments that increase intensity *do not even offer* the *possibility of* containing the material impact of production. This conclusion is restricted, of course, to *this* specific society, the capitalist one, in which the growth of production is an unavoidable imperative.

Struggling to visualise the practical difference between a gain in productivity and a gain in intensity is not unusual. A very simple and schematic example will help visualising the difference.

Suppose a production conveyor belt is introduced to a workshop where artisans work at a given pace. However, the *pace* of work does not change; it only changes its characteristics. This corresponds to an increase in productivity. However, if this same conveyor belt starts being programmed so that it can be accelerated, the pace of work will have to increase, corresponding to an increase in intensity. In any case, it is absolutely common for changes in the production process to bring about both effects. Even so, this division allows for the approaching of objects from different angles and the observation of some significant differences from which meaningful comparisons may be derived.

Productivity gains cause the number of units produced to increase, but the *magnitude of the value* produced in a given period remains constant. The value of labour-power decreases, which increases surplus value. In other words, the distribution of the new value created by labour changes. A smaller portion is now sufficient to remunerate the worker, which implies that the portion that remains as surplus is now larger. One part of the new value decreases so that the other can increase.

In the case of intensity gains, the overwork compressed into the working day is equivalent to the regular work done over a longer working day. Therefore, it creates value *as if* the working day had, in fact, been extended. In this case, contrary to what happens when there are productivity gains, surplus labour time increases even without a reduction in the necessary labour time. Both the new value and the surplus value grow because the amount of surplus labour has grown.

To sum up, with higher productivity there will be a larger amount of products, the same amount of labour for a given day, and the same value is distributed over a greater number of products, which also means a reduction in the value of each unit produced. With higher intensity, there will be a larger amount of products and a greater expenditure of labour on a given working day. Assuming that the increase in intensity is general, as done in the case of productivity, the result in terms of the sum of the value of the total product is the same.

The point to emphasise is that if the length of the working day remains unaltered, raising productivity and intensity are both ways of expanding surplus value. In the first scenario, this is achieved in the form of relative surplus value; in the second, in the form of absolute surplus value. This will clarify the way productivity and intensity relate to efficiency.

It has been demonstrated that the expansion of relative surplus value depends on a decrease in the value of labour-power. A trivial example may illustrate this. Today, the importance of energy and electricity consumption in workers' lives is enormous. Therefore, increasing energy efficiency also reduces the relative costs of energy consumption for the worker. Increasing energy efficiency is, therefore, a way of expanding relative surplus value. In this sense, increasing efficiency fulfils the same objective associated with increasing productivity.

At the same time, the expansion of absolute surplus value through increased intensity requires an acceleration in the pace of production. This corresponds not only to an intensification of labour but also to an intensification of the use of the means of production. As a result, the means of production tend to wear out more quickly. However, if efficiency is increased, the rate at which the means of production process inputs increases less quickly than the rate at which productive activity increases. In this way, the wear and tear on the means of production is *relatively* reduced. In this sense, efficiency gains make it technically feasible to increase intensity. It, therefore, enables a specific way of expanding surplus value.

Finally, it is necessary to tackle the greatest trump card of the dematerialisation argument. Indeed, raising the efficiency level brings the possibility of saving material and energy resources in production. A more efficient production process achieves the same result by consuming fewer raw and auxiliary materials. Therefore, the question is: What if this promise of saving material resources were fulfilled? Would it offset the other deleterious effects? Taking this possibility to its limit, couldn't the fable of the dematerialisation of production become a reality and thus allow for an ethereal and sustainable capitalism?

Considering the theoretical elements presented here, it is safe to answer that no, this possibility is not realisable under capitalism. The reason is surprisingly simple. If the possibility of saving resources brought about by greater efficiency were realised, the material impact of this production would decrease. However, these saved resources are not only physical things. They are also capital. When these resources are saved, capital previously advanced through them no longer needs to be advanced. In other words, reducing material consumption in production also means saving capital. In other words, part of the capital that previously operated there is released. As previously mentioned, capital cannot settle into immobility. Immobility directly violates its essence. Therefore, this liberated capital must, at all costs, find other opportunities to carry out its expansion logic. When it does, it necessarily re-establishes contact with materiality.

The conclusion is that, in the very nature of capital, there is a mechanism that neutralises even the narrow possibilities of moderating the material ecological impact of capitalist production.

The final part of this chapter takes a deeper approach to the discussion of the mechanisms underlying the large and complex network of events that is the development of the productive forces.

The development of productive forces is one of the most critical dimensions in capitalist society dynamics. As seen before, productive forces develop through technological and organisational changes. This process can also be understood as the rise in efficiency, productivity, and intensity levels.

Demarcating the difference between efficiency and productivity is of utmost importance. The first saves resources, but only in relative terms. The second necessarily increases material consumption in production. There are also relationships that can be established between these three dimensions. Efficiency and productivity combine to expand relative surplus value, and efficiency can be a technical condition that enables the expansion of absolute surplus value via an increase in intensity.

This is the theoretical foundation for a deeper discussion on the advance of productive forces, trying to understand how the possibility of appropriating a greater mass of value drives capitalists to develop the productive forces under their command. However, "the possibility of *appropriating* value" differs from the discussion presented until now, which refers to the increased *creation of* value and surplus value.

To begin with, we need to return to a familiar category: value. As stressed a few times, the value of a commodity is determined by the socially necessary labour time to produce it. The crucial word here is *socially*. It is not the labour time directly employed in producing a given commodity that determines its value. In other words, it is not the labour employed in each production process.

This socially necessary time is determined by the average conditions of production in the specific branch of each commodity. This is important because when capitalists sell commodities for their value, they *do not* receive the equivalent of the labour directly expended in their production. They receive the equivalent of the *average social labour* time objectified in those commodities. There is a difference between the time directly employed and the socially necessary time, which is very important for the rest of the argument.[3]

What happens if the labour time directly applied is less than the socially necessary labour time? Capitalists take commodities to the market and realise their value, i.e., sell commodities for their value. In doing so, they appropriate the magnitude of value corresponding to the average social labour. In other words, they appropriate labour that they did not actually employ. Conversely, if the labour directly applied is greater than the average social labour, capitalists appropriate the magnitude of value corresponding to the average social labour as in the previous case. However, in this case, the objectified labour appropriated is less than the amount of labour carried out at their command.

It is possible to predict where this argument is going. Supposing that the labour time directly applied is equal to the socially necessary labour time, capitalists go to the market and appropriate the exact equivalent of the labour performed at their command. In this scenario, what happens if a capitalist introduces a technical improvement? Or an improvement that increases productivity? If the productivity of a specific production process increases, the result is that the same amount of directly applied labour will produce a greater number of units in a given amount of time. As a result, the amount of labour directly applied to each unit decreases

But what about the value of these units produced? If the market is large enough, the value would either stay the same or change very little. Even if this capitalist introducing a technical improvement has significant weight within the market, the value of the commodity would decrease, but not in the same proportion as the labour time directly applied to this production of a higher level of productivity.

Because this is an especially challenging part of the argument, it is worth repeating it another way. In the previous part of this chapter on technology, the reasoning was constructed with reference to the general increase in productive forces, i.e., the increase in the *general* level of productivity, efficiency, and intensity. The present approach looks at the issue from a different perspective. In this part, the focus is on *isolated* variations in productivity. As these variations are isolated, they have little or no effect on average levels. And since the value is determined by the socially necessary labour time, which in turn is determined by the average production conditions, isolated variations in productivity have little or no effect on the value of the commodity.

Having said that, the previous example shows that the capitalist who operates at a level of productivity above the average needs to apply less labour to obtain the same result as any competitors, i.e., to realise the value of the commodity by selling it.

In this case, the increase in productivity creates a difference between average social labour and directly applied labour. The capitalist appropriates more objectified labour time than the labour performed at their command. This difference is what Marx calls extra surplus value. Since it is precisely the pioneering technical improvement that allows the capitalist to appropriate this additional value, there is a structural drive to always strive for higher levels of productivity.

It is worth mentioning here an aspect of this discussion relating to consumption, which Chapter 6 will explore in greater depth. Notice that an increase in productivity results in more units being produced. However, it is not enough for them to be produced. These units must be taken to the market *and sold.* Otherwise, the value is not realised, and neither is the extra surplus value.

Resuming the thread of this presentation, it has been mentioned that this possibility of increased appropriation of value exists and that, to realise this possibility, a capitalist needs to achieve a higher-than-average level of productivity. The question now is: what are the other competitors expected to do? They will endeavour to do precisely the same.

And what happens as a growing number of capitalists raise the level of productivity in their respective production processes, thereby reducing the amount of labour they directly apply? If this reduction in directly applied labour is sufficiently widespread, the *socially* necessary labour time tends to decrease. This is because the average conditions of production are becoming more productive. With this universalisation of higher levels of productivity, the gap that previously existed between the labour time directly applied by some capitalists and socially necessary labour time tends to disappear. Thus, considering this process alone and ignoring other disturbing influences, the extra surplus value tends to dissipate.

But there is more to it. With the generalisation of the most productive techniques, the value of the commodity is reduced. This has a retroactive effect on any producers who have not yet, for whatever reason, adopted the best techniques. Any laggard capitalists using the old methods will be compelled (forced) to adopt the new techniques. In this case, not to gain advantages but to try to minimise disadvantages. Even those who used to operate at the average level will now be below average if they fail to make changes. If they keep applying the same amount of labour as before, they will now appropriate less value, since the average conditions have moved. They are, therefore, obliged to adjust, regardless of their will. *Their freedom not to do so is merely formal.*

In short, the whole process is driven by extra surplus value. Capitalists who get ahead appropriate more value. In the wake of this, there tends to be a process of generalisation of the best techniques as competitors seek to obtain the same advantages. This generalisation gradually dissipates the extra surplus value, the advantage that triggers the whole movement. In the process, the value of the commodity decreases. When this happens, even the laggard capitalists are forced to make changes. All these articulated developments point to a very vigorous self-reproductive logic of raising the productive forces.

The exact same reasoning applies to discovering another value differential that also drives this logic. It has been said that value is determined by the socially necessary labour time. So far, with the exception of directly applied living labour itself, all the conditions of production were kept constant as average conditions. In other words, in the previous exercise, the variations explored only regarded the need to apply living labour time.

However, the law that determines value according to labour time does not refer exclusively to living labour. It applies to labour as a whole, including labour objectified in the means of production. And the labour objectified in the means of production, like living labour, also makes up the value of the commodity as average social labour. Therefore, if the conditions of production relating to the means of production are applied at the average social level, then the value transferred to the commodity corresponds precisely to the value that existed in them before.

There is another way to demonstrate this relationship. The means of production transfer value to the commodity, i.e., they transfer a portion of their value as consumed in the production process. However, their value is determined by the labour time that is socially necessary to produce them. And the value they transfer is determined by the average conditions of production in the sector in which they are employed. Considering this is a complex discussion, this exploration will rely on reinforcement and repetition as it advances.

If the average conditions of production also determine the value transferred to commodities, a particular type of means of production, raw materials and auxiliary materials, must be assessed. Their specificity is that they are entirely consumed in each production process. Thus, they transfer all their value in each production process.

Now imagine a gain in efficiency. Greater efficiency means a lower ratio between input and output. In other words, it is possible to obtain the same level of product by consuming a smaller quantity of raw and auxiliary materials. However, if there have been no changes to the processes in which these inputs are produced, their value remains unchanged. More importantly, the value transferred to the end product does not change either.

The logic here remains the same as the one applied to productivity. The value transferred does not change because the efficiency gains only occur for an individual capitalist. In terms of the average conditions of production, the ratio between input and output remains the same. And since the share of constant capital that makes up the value of the commodity is determined by the average conditions of production, the value of the commodity remains, at least initially, the same.

As a result, just as in the previous relationship, a difference is created between the labour directly applied in production and the average social labour appropriated from the sale of goods. The higher the level of efficiency in relation to the average, the greater this differential in value suitable to appropriation.

Therefore, in the same way that it is possible to demonstrate an inherent drive to gain productivity, it is also possible to demonstrate an inherent drive to gain efficiency. And, against all the hopes placed on the technological prowess of capitalism, it is clear that productivity gains do not even have the possibility of moderating the material ecological impact of production. It is clear that, despite the saving possibility of greater efficiency, this possibility will never be realised in capitalism. Chapter 7, on market mechanisms, explores this conclusion in greater depth.

To close this chapter, recognising the difficulty of keeping track of so many details, a summary of its main points is presented below. In the chapter on growth, the conclusions were highlighted point by point. This was because they essentially formed a single conclusion: in capitalism, production grows out of necessity. It grows as value. It grows as use value. It grows materially. In this chapter on technology, this procedure does not apply because the conclusions are many, varied, and interconnected in a non-linear sequence of arguments.

Hence, it will be more effective to return to the beginning and demarcate the conclusions against the ideas to be demystified. Contemporary environmental policy and its advocates bet on dematerialisation. Faith in dematerialisation is based on the input-saving potential of new technologies that increase production efficiency.

At this point, the initial discussion has already led to a critique: this dominant view overestimates a single dimension of the advance of productive forces. However, in addition to efficiency gains, there might be developments in productivity and intensity that bring an increase in productive consumption.

In other words, productivity and intensity gains not only frustrate expectations of dematerialisation but also push global production in the diametrically opposite direction. Nevertheless, more important than pointing out possible limits to dematerialisation is to demonstrate that these are not just possible, casual, and conjunctural limits but *systemic*, structural ones. In other words, to demonstrate that there are mechanisms that continually push and renew this dynamic, out of necessity.

And what is the goal behind all the processes that raise the productive forces analysed so far? All the reasoning presented here is based on the categories of value and surplus value. Of course, it is a well-known fact that what motivates the conscious actions of capitalists is profit. Profit and surplus value are not equivalent categories. It is unusual for people to think about surplus value. But the discussion of surplus value allows us to explain what happens so that profit responds, at least initially, in the way intended by those who implement technical modifications. Here, only the mechanism by which surplus value suitable to appropriation expands is explained. Later, with other mediations and determinations, this mechanism will be part of the explanation of why profit expands.

In conclusion, capital achieves the expansion of surplus value in four ways. First, by contracting the labour time necessary for the reproduction of the labour-power by raising the general level of productivity. As a result, surplus labour time is extended, increasing relative surplus value. Second, by creating a differential between directly applied labour and socially necessary labour. This is achieved by increasing productivity and/or efficiency above average levels. Third, by compressing the work that would be done in a normal day into ever shorter periods, through greater intensity imposed on the labour process. Fourth, by the imperative elimination of waste in the production process. This fourth point may initially seem to be a virtue of the blind logic of capital, but when market mechanisms are discussed in Chapter 7, it will become clear that this is not the case.

Why insist that these movements occur out of necessity? Because a capitalist's freedom to submit or not to these series of impulses is merely formal. Their individual will is constrained by the competition between capitals. If the owners of capital do not respond to these impulses, and if they do not continuously seek to raise the productive forces under their command (as pioneers or laggards), their capitals run the risk of being destroyed or absorbed by larger ones. If they do not act "accordingly," they are exposed to the risk of losing their capitalist status through competition. Consequently, impulses become compulsions in an imperative way. In Marx's own words:

> The law of the determination of value by labour-time makes itself felt to the individual capitalist who applies the new method of production by compelling him to sell his goods under their social value; this same law, acting as a coercive law of competition, forces his competitors to adopt the new method.[4]

Finally, what conclusions can be drawn in terms of ecological implications? The dynamics of the development of productive forces and the material impact of our productive activity are, taken as a whole, outcomes of the same cause in the present social formation. A cause emphasised many times in this book: production governed by value. What we take from this is that efficiency cannot cancel out (or counteract) the increase in consumption. Furthermore, the types of causality actually established between efficiency and consumption always go in the direction of reinforcing one and the other. On the one hand, consumption can only grow from a previous development in efficiency. On the other hand, the technical development that promotes efficiency is only economically viable if it can cause capital to grow. If all these constraints, all these imperatives, and all these combined results are specific to the social metabolism of capitalist society, then the only possible conclusion is that capitalism is not ecologically viable. Therefore, overcoming capitalism is a prerequisite for any claim to achieve sustainability.

Notes

1 Smith, *The wealth of nations* (2003).
2 Marx, *Capital*, vol. 1 (1982), 129.
3 For a more detailed categorical analysis of this excerpt, see Sá Barreto, *De Smith a Marx: curso introdutório em dez aulas ["From Smith to Marx: an introductory course in ten lessons"]* (2021), Lesson 10.
4 Marx, *Capital*, vol. 1 (1982), 436.

References

Marx, K. (1982). *Capital* (Vol. 1). London: Penguin Books.
Sá Barreto, E. (2021). *De Smith a Marx: curso introdutório em dez aulas*. São Paulo: Usina Editorial.
Smith, A. (2003). *The wealth of nations*. New York: Bantam Classic.

5 Environmental ethics and the question of the objectivity of values

The first chapter examined how Ecological Economics pays special attention to consumption patterns. The general equation, so to speak, is that current consumption patterns are wasteful and, therefore, push humanity as a whole into destructive interaction with the environment. Consequently, it is assumed that if consumption patterns were different, guided by a higher ecological awareness, technological advances would create the possibility of an affluent society with a low material impact.

Indeed, it would be unfair to reduce Ecological Economics to this kind of expectation (which must now seem rather naïve). At this point, what matters is to remember that this kind of expectation is, in fact, part of the interpretative structure of this tradition. More importantly, it is essential to bear in mind that such reasoning, which is enunciated with varying levels of sophistication by ecological economists, can also be found very frequently in less formal discourses. In other words, this understanding of consumption patterns is already part of common sense.

From the most liberal ones, which focus on individual change, to the most utopian ones, which focus on a change in universal consciousness, a wide range of perspectives share the more or less well-defined idea that consumption patterns need to be reorientated (individually or collectively). The problem to start discussing in this chapter does *not* lie in this general conclusion. It is, of course, a correct conclusion. Indeed, if the interaction between humanity and nature is marked, among other things, by productive and unproductive consumption, it is clear (almost by definition) that unsustainable consumption keeps us from any ambition of ecological sustainability. A sustainable society would, therefore, presuppose overcoming these consumption patterns, as even common sense would state. The problem lies in the means by which this overcoming would be achieved.

The most liberal perspective rests on a well-known soundbite: "If everyone does their bit...." The conclusion of this formula, in which the expected result would finally be realised, is always thwarted. And the culprit is always the individual who didn't do their bit. It is not enough to recognise the absurdity of

DOI: 10.4324/9781003512516-6

this formula. It must be demonstrated that there are objective limits even to the premise "if everyone does their bit." In other words, to expose why the generalisation of virtuous behaviour, based on individual decisions, is hindered on the one hand and compensated on the other. The next chapter is devoted to this endeavour.

The most utopian perspective seems superior at first since it does not fall into the trap of individual responsibility. Instead, it calls for universal awareness, i.e., a new environmental ethic. However, perhaps even more than the previous perspective, this one is also systematically thwarted at its roots. Again, it is not enough to note that the universal adoption of a new ethic (abstractly conceived) is implausible. It must be demonstrated that there are objective limits to this subjective turn.

Its fundamental weakness is the translation of environmental concerns into abstract ethical precepts. This is a weakness because the desired new ethics is simply postulated. Its relationship to objective presuppositions of social practice is not investigated. To fill this gap, it must be clarified that the values that make up any ethics *are not purely subjective determinations*. In other words, values are *also* objective determinations.

Hence, one must enquire what are the conditions for their realisation. Strictly speaking, any ethical principle can be formulated, including those aimed at ecological sustainability. However, not every abstractly formulated ethical principle is realisable. It is possible that the social conditions in place will sometimes create obstacles to their realisation, sometimes make them completely impossible, and sometimes compel individuals to adhere to practices that conflict with the values they profess.

More than simply emphasising this possibility, Medeiros and Sá Barreto seek to demonstrate that:

> in capitalist society, all values, including those presupposed by "environmental ethics" [...], have their realisation subordinated to a kind of objective ethics: the ethics of capital. This means that the theoretical problem of uncritical environmentalism takes on a practical expression: the theoretical error of postulating values becomes a problem of realisation because the values postulated are incompatible with the values that characterise capitalist society on an ethical-moral level.[1]

To put it very bluntly, the postulated values are irreconcilable with the reproduction of capital and, therefore, practical everyday reality ends up demanding from individuals a behaviour incompatible with the postulated values, no matter how firmly individuals are convinced of the legitimacy of these values.

It is easy to see that maintaining this conclusion is not a trivial task. That is why this first part is divided into two parts. First, a deep exploration of the relationship between values, ethics, morality, and practice. Then, a demonstration of how the so-called ethics of capital *objectively* subordinates any other values.

The contributions of Lukács are a good start. This Hungarian Marxist philosopher sought to recover the ontological character of Marxian thought precisely to develop a materialist reflection on ethics. The Hungarian thinker died before completing his philosophical project, but its first steps may already take this debate a long way.

In his later work, *Ontology of Social Being*, Lukács recovers from Marx a formidable *insight* into the fundamental trait distinguishing the human species from other animal species. According to the author, this trait is our unique ability to perceive possibilities in the world that have not yet been realised.[2] In other words, it is the ability to perceive reality not only as containing everything that is already in place, i.e., already realised, but also as containing a series of things that *can be*.

This peculiarity in our way of perceiving the world has decisive repercussions on the nature of the typically human productive activity, hereafter called *labour*. The fact that humans can perceive what *not yet is* but *can be* enables us to seek the fulfilment of our needs with this overload of possibilities in the world. This is why labour is marked by intentionality. In other words, having recognised the possibility of bringing something into the world better suited to meeting a specific need, humans take the realisation of this possibility as the purpose of their actions. In *Capital*, we find a passage that emblematically presents this reasoning:

> A spider conducts operations which resemble those of the weaver, and a bee would put many a human architect to shame by the construction of its honeycomb cells. But what distinguishes the worst architect from the best of bees is that the architect builds the cell in his mind before he constructs it in wax. At the end of every labour process, a result emerges which had already been conceived by the worker at the beginning, hence already existed ideally.[3]

Therefore, not only does the purpose motivate the action, but it also presides over the entire process as a kind of guiding principle. In other words, it *is essential that the conduct undertaken is compatible with the realisation of the* purpose. But why is this important to understand the relationship between values, ethics, morals, and practice? According to Medeiros and Sá Barreto:

> it is because labour is an activity aimed at achieving a previously defined goal that not only does this goal emerge as the guiding principle for the entire process of objectification [...] but also as the basis for the behaviour of those who labour [...] and as a criterion for judging the adequacy of a practice.[4]

Untangling this sentence, by recognising that "labour is an activity aimed at achieving a previously defined goal," one recognises the objective foundation of values. By stating that this "goal emerges as the guiding principle of the entire process of objectification," one identifies the objective foundation of ethics. By asserting that, as a guiding principle, the goal conditions the behaviour of those who work, one finds the objective foundation of morality. Finally, by recognising that goal is compared to the results of a practice, as a criterion for evaluation, one recognises the objective foundation of value judgements.

When challenging the overvaluation of subjectivity in the treatment of ethics in general, this chapter consequently challenges the so-called environmental ethics. However, following a completely opposite direction may lead to a different version of the same mistake. Recognising that values have an objective foundation *does not mean denying the subjective dimension* entirely. What it does imply is recognising that subjectivity *is not completely free*. In other words, subjectivity, whatever it may be, always has objective roots, on the one hand, and always faces objective conditions of realisation, on the other.

This second aspect, related to the conditions of realisation, is crucial. If values (and therefore also ethics, morality, and value judgements) always have an objective foundation, then they always have *a certain level of* contact with reality. If that were all there was to it, the initial statement that the values of environmental ethics are abstractly conceived or simply postulated would be highly questionable. But there is more to it. This section illustrated how values originate in the finalistic nature of human activity. Also, that this human peculiarity stems from the fact that we perceive possibilities in the world that have not yet been realised. The short commentary on morality and value judgments highlighted how the adequate realisation of a goal requires a choice of methods and courses of action that are compatible with it. In conclusion, even those values that arise in a given era may not find the objective conditions for their realisation in that same era.

Therefore, Medeiros and Sá Barreto state:

> the values of uncritical environmentalism also have an objective foundation. As legitimate values emerging from contemporary social coexistence, they express, on the level of moral consciousness, the social tragedy that contradictorily accompanies the human conquest of nature in capitalist society. The question, therefore, is not whether or not these values are objective – which we answer in the affirmative. The question is whether or not the objectivity of these values is based on an objective contradiction that makes their realisation unfeasible at the present time.[5]

Again, this unfeasibility will be demonstrated in the next chapter. First, this chapter concludes by addressing the second of the two previously mentioned moments.

After discussing the relationship between values, ethics, morality, and practice, the next issue to be approached concerns the conditions for realising values and, still at a relatively high level of abstraction, how the ethics of capital subordinates all other values.

So as not to overextend this section, a slightly different path to the one found in Medeiros and Sá Barreto's article is taken here, favouring a review of the discussion of commodity fetishism.[6] The choice of fetishism has an interesting advantage. It is very common, in hasty readings, to take this well-known passage from Marx's main work as a kind of moral critique of capitalism. This kind of reading usually departs from the famous statement that, in capitalism, the relationship between people appears as a relationship between things. It concludes that this state of affairs originates from an obsession with commodities instilled in people or developed from some fixation or exaggerated material attachment. The corollary of such a reading is that it would be enough to identify this illusion and elucidate its origin in perversion or moral weakness. As a result, fetishism would somehow dissolve into other, more virtuous ways of relating to each other and to the products of human labour.

A critique like this, which clearly overestimates subjective aspects, may as well be made. However, it does not carry the full weight of the Marxian analysis of fetishism. The brief contact with Marx's theory of value offered in the previous chapters provides enough elements to affirm that, in practice, what matters to the agents of exchange is how many other people's products they can get in exchange for their own. This importance does not stem from some ill-informed personal preference. It stems from the very objective fact that, in a commodity-producing society, almost everything necessary to live must be extracted from the market, and the quantity and quality of what one extracts from the market is determined by the proportions in which products are exchanged.

Such proportions, however, vary constantly, independent of the will, foresight, or actions taken by those carrying out the exchange. What defines the proportions in which commodities are exchanged is their respective magnitudes of value. And their magnitudes of value, in turn, are determined by the social conditions of production in which the individual producer plays only an atomised role. Therefore, the magnitude of value is determined by the *socially* necessary labour time.

This is why the social movement of the products of labour and the proportions in which they are exchanged occurs, in fact, as a movement of things, escaping the will and control of the producers. In fact, it is they who are subordinated to this movement. Subsequently, this result may be translated into more commonly used language by saying that there are already enough elements to affirm that the subjects are subordinated to the oscillations and moods of the market. This is a critical conclusion of significant weight.

This is what Marx demonstrates in section 4 of Chapter 1 of *Capital*. He shows that, in capitalism, human labour appears as a property *of commodities*, i.e., as objectified labour, as value. Also, that the measure of the expenditure of labour appears as the magnitude of the value *of commodities*. And that the social character of labour appears as a relationship between *commodities*. Notice how, in these last three sentences, it is clear that the typical traits of human activity are reflected as a property of *the products of* that activity.

In conclusion, the relations between commodities are determined by their respective magnitudes of value, which are determined by the socially necessary labour time. This time, in turn, is determined by the current normal conditions of production. Thus, the circumstances in which producers participate in exchange are determined by conditions completely beyond their control. The fluctuations in the average conditions of production occur regardless of the producers' intention, will, and direct action. This is because, in Marx's words, "in the midst of the accidental and ever-fluctuating exchange relations between the products, the labour-time socially necessary to produce them asserts itself as a regulative law of nature."[7] That is why it dominates them, subordinates them. And it is not controlled by them.

Even through this very brief recapitulation, it is already possible to conclude that people's entire material lives are subordinated to how they manage to participate in the market. The more successful an individual is at navigating market forces, the more secure and affluent their material life will be. The more the individual's behaviour brings them into conflict with these forces, the more their primary living conditions are put at risk.

So, it is no surprise that the urgencies of everyday life systematically require individuals to behave in a way that suits this commodity-producing structure. Strictly speaking, a behaviour suited to the reproduction of capital. Nor should it surprise if these behaviours often conflict with the values these same individuals profess, including those that would be part of an ethic oriented towards environmental sustainability.

According to Medeiros and Sá Barreto,

> the proposals analysed heavily rely on advocating the conscious regulation of capitalist production, resource use, and private consumption. This regulation would be achieved by exhorting and mobilising individual practices in an environmentally "appropriate" direction. It is true, on the one hand, that these proposals play an important role in exposing current patterns of production, consumption, pollution, and environmental degradation. On the other hand, there is a clear general understanding that reversing these trends would occur through ethical and moral transformation. The defence of a new ethic, therefore, is based on the belief that production, although subject to capitalist

logic, would be subordinated to the designs of a *radically renewed*, ecologically responsible conscience committed to environmental sustainability.[8]

The next chapter will demonstrate the various ways in which this belief is mistaken.

Notes

1 Medeiros & Sá Barreto, "Lukács e Marx contra o 'ecologismo acrítico': por uma ética ambiental materialista" ["Lukács and Marx against 'uncritical ecologism': for a materialist environmental ethic"] (2013), 318.
2 Lukács, *Ontologia do Ser Social I* [*Ontology of social being*, vol. 1] (2012).
3 Marx, *Capital*, vol. 1 (1982), 284.
4 Medeiros & Sá Barreto, "Lukács e Marx contra o 'ecologismo acrítico': por uma ética ambiental materialista" ["Lukács and Marx against 'uncritical ecologism': for a materialist environmental ethic"] (2013), 324.
5 Ibid., 327.
6 If this part is not detailed enough, it is possible to find a less telegraphic exposition in Sá Barreto, *De Smith a Marx: curso introdutório em dez aulas ["From Smith to Marx: an introductory course in ten lessons"]* (2021), Lesson 08.
7 Marx, *Capital*, vol. 1 (1982), 168.
8 Medeiros & Sá Barreto, "Lukács e Marx contra o 'ecologismo acrítico': por uma ética ambiental materialista" ["Lukács and Marx against 'uncritical ecologism': for a materialist environmental ethic"] (2013), 323.

References

Lukács, G. (2012). *Para uma ontologia do ser social I.* São Paulo: Boitempo.
Marx, K. (1982). *Capital* (Vol. 1). London: Penguin Books.
Medeiros, J., & Sá Barreto, E. (2013). Lukács e Marx contra o "ecologismo acrítico": por uma ética ambiental materialista. *Economia e Sociedade, 48*(2), 317–333.
Sá Barreto, E. (2021). *De Smith a Marx: curso introdutório em dez aulas.* São Paulo: Usina Editorial.

6 Consumption and the myth of individual moderation

Chapters 3 and 4 discussed the expansion-oriented impetus of capital in terms of the unstoppable growth of production and the advance of the productive forces. In conclusion, it was found that a capitalist society cannot be anything other than expansion-oriented. And also that, under the logic of capital, technology cannot be anything other than a booster of material impact on the planet. Throughout that exposition, despite the constant mention of the need for product realisation, i.e., the sale of commodities on the market, the focus of the analysis concentrated on the sphere of production.

Therefore, some fundamental peculiarities of the moment of commodity circulation in the process of capital valorisation have yet to be considered. This chapter broadens the scope of the debate to include important details of the entire movement of capital circulation.

When considering its complete cycle, the valorisation of capital is also determined by the *circulation of commodities*, which is external to the sphere of value creation. In other words, from now on, the investigation focuses on the moments when capital is *not* in the sphere of production. This happens when it appears in the form of money buying means of production and labour-power, or in the form of commodities, already increased in value, seeking to realise itself through exchange. In doing so, the requirements imposed on consumption during the movement of the capitalist mode of production will be addressed.

The first step is a simple characterisation of the capital cycle. According to Marx, capital is a unity of three processes (therefore, if capital *is* this unity of three processes, it *only exists* in this unity). The first process is that in which capital leaves its initial money form to take on the form of means of production and labour-power. In the *Grundrisse*, Marx calls this first-stage "devaluation".[1] This is not devaluation as such because there is no reduction in the initial value; it is just a change of form. Marx names it this way simply because value is leaving its autonomous form of value, money. The second process is valorisation in production, which is everything analysed in detail so far. It consists of the expansion of capital through its transformation into commodities increased

DOI: 10.4324/9781003512516-7

in value. The third process is valorisation through exchange. Again, the term "valorisation" does not indicate an expansion of value in exchange. This third moment does not include any creation or expansion of value. The commodity produced, which now goes to the market, is already a sum of value increased in relation to the sum that started the movement. But in exchange comes the necessary, unavoidable moment of its *realisation*. The moment when the increased value takes the autonomous form of value, money. In other words, throughout the production process, capital is reproduced and expanded, but in the form of commodity. To resume its cyclical movement, capital must necessarily return to the form of money, which requires the sale of the commodity produced.

From the point of view of the reproduction of capital, a challenge arises: despite existing only through this unity of devaluation, valorisation in production and valorisation through exchange, the internal cohesion between these three moments can be fulfilled or not. There is no automatic fluidity in this alternation of forms. There is no guarantee that they will actually occur.

In other words, although they form a cyclical movement that must be continually renewed, these three moments exist in relative autonomy. Thus, each moment does not, on its own, set the conditions that guarantee the successful passage to the next moment. The unity of the three processes, the very thing that constitutes capital in motion, has to be constantly affirmed in practice. And it is precisely the requirements on consumption created by this necessity to be presented next, discussing four types of requirement: scale, time, scope, and speed.[2]

The first requirement discussed here is scale. The demonstration of this first dimension is actually quite straightforward. Discussions developed in the previous chapters concluded that the internal logic of the capital valorisation process has the necessary result of expanding the number of commodities released into circulation. Also, the existence of capital is not limited to the moment of valorisation; its existence can only be affirmed by the continuous alternation of forms in a cycle that cannot be interrupted.

These two elements must come together. At a certain moment, the internal logic of capital results in an expansion of production. To exist as capital, the sum of the value of the commodities must be fully realised, i.e., the product must be completely sold. From the point of view of capital valorisation, it does not matter whether or how the product sold is consumed. But even so, it can be assumed with reasonable certainty that to guarantee a growing scale of sales, it is also necessary to ensure an ever-increasing scale of consumption.

Certainly, each of us is individually capable of moderating our consumption. The limit for reducing consumption does not operate *at the individual level*. Nonetheless, capital can only guarantee the conditions for its healthy reproduction (i.e., permanent growth) if it can ensure that the *total* scale of consumption is also growing. From an ecological point of view, this is the benchmark that matters. Our ecological impact on the planet is determined by the total scale of our activities.

There is another, more indirect way of making the same point. It was stated that the commodity is a unity of value and use value, as taught by Marx. Also, that there is no causal link between use value and value. In other words, "more useful" objects do not necessarily have a greater magnitude of value. On the other hand, the existence of any use value is a condition for the existence of value. An object without use value cannot have value, even if it is a product of human labour.

The big problem for the process of valorisation of capital is that use value, unlike value, does not have an unlimited character. If production, in its use value dimension, exceeds society's consumption capacity at a given moment, the surplus mass of commodities ceases (at this limit) to be use value. It is essential to pay special attention to this part of the reasoning. Excess commodities may well have useful physical properties, but if there are no more needs to be met through them, they have no use value. And if they have no use value, they have no value.

Therefore, according to Marx, capital finds a limit in other people's consumption. A limit that it must continually and systematically overcome. A limit that is overcome by an expansion-oriented scale of consumption, by ensuring that society's capacity to consume grows repeatedly.

The discussion on the requirement related to time must be divided in two parts. The first concerns the necessary continuity and fluidity of the circulation of capital. The second concerns the necessary renewal (often artificial) of the need that motivates consumption.

To some extent, the first part is already contained in the previous discussion, but it can be further elaborated. As seen before, capital must go through its cycle, successively switching between forms. It begins as money, which is converted into means of production and labour-power, which are then transformed into commodities, which are then reconverted into money. Or, to put it in other words, a sequence of devaluation, valorisation, and realisation. Or, again, an alternation between circulation, production, and circulation.

As highlighted, capital, in order to be capital, must continually move from one form to another. But there is another approach to the matter. As long as capital remains in production, it cannot circulate. In other words, when it is in the form of means of production or labour-power, it cannot realise the value contained therein. Nor, to some extent, can it be applied in other branches of production. Similarly, while capital remains in circulation, it cannot produce. Thus, it cannot absorb surplus labour, surplus value. As a consequence, it cannot expand. Furthermore, it is in commodity form and therefore at imminent risk of not having its value confirmed by the total consumption of society.

Furthermore, there is a necessary complement to the previous statement that capital needs to alternate successively between its various forms. Capital only takes on another form by negating the immediately preceding form. When undergoing the production process, it expands as capital but cannot yet be realised. In circulation, it finds its way back to the money form but cannot be expanded. This clarifies the possibilities and *impossibilities of* capital at each stage of its cycle, or in each of its forms.

This intermittency, however, is not observed in practical terms. As Marx warns, capital does not carry out these alternations of form in full. What actually happens is that, at all times, there are fractions of capital present in each of its forms. In other words, when a portion of capital leaves the commodity form to take on the money form, another portion must leave the form of productive capital to take on the commodity form. This happens with each phase, at a pace determined by the conditions of production. In short, there must be a certain fraction of capital in each of its forms at each moment. In this way, each stage of the cycle guarantees its prerequisite and, at the same time, constitutes the set of conditions for the next stage.

The change of form must, therefore, not only be continuous. It must occur in a roughly simultaneous (synchronised) way at all points of the cycle. If the movement of capital stalls in one of the forms, this coordinated fluidity of the whole process is undone. Capital excessively accumulates at one stage and becomes scarce at others. Also, the adequate proportion between the different shares of capital at each stage is broken. In short, it is only through the unity of the three processes mentioned above *and the continuous fluidity* of the transition of the shares of capital between the phases of this cycle that, according to Marx, "the continuity of the overall process" is accomplished.

As mentioned before, when approaching the requirement related to time, the discussion must be divided into two parts. This was the first, about the continuity and fluidity of the circulation of capital. The second part concerns the necessary renewal of the needs that drive consumption.

The entirety of the capital advanced only completes its cyclical movement when all the initial value has finished passing through all of its forms, returning to the money form. The time needed for capital to fully go through all these stages is called the *turnover of capital*.[3] Throughout a complete turnover, the capitalist has to advance capital before the moment arrives to recover it in its entirety in the money form (and in expanded magnitude).

This category will be important later on. First, there are two new categories to be introduced. It has been mentioned before that variable capital is the fraction of capital converted into labour-power, and constant capital is the fraction converted into means of production. The following paragraphs will present *fixed capital* and *circulating capital*. It is imperative not to confuse these four categories. Considering that capital is divided into constant and variable, how these elements of capital move through the valorisation cycle will determine to what extent they are fixed or variable capital.

The insight to be retrieved from Marx is: if, at any stage of the cycle, the capital in question is *value*, it is circulating capital; and if the capital is involved in the valorisation process, but *only as use value*, it is fixed capital.[4]

To clarify this reasoning, it is helpful to break it into smaller parts. Labour-power is variable capital. At *each production process*, its value is destroyed through the workers' consumption and restored through the act of labour, which creates new value. At each production process, therefore, the entire value of labour-power participates in the valorisation cycle. For this reason, variable capital is always also circulating capital. But that is not what matters most to this argument. Let's look at constant capital.

The elements of constant capital can be divided into two groups. Raw materials and auxiliary materials, on the one hand, and equipment, machinery, infrastructure, etc., on the other. As mentioned before, the first group fully transfers its value to each production process. Therefore, it participates completely as value in each process. As such, this first group is circulating capital in its entirety. As also mentioned, the second group gradually transfers its value to the product throughout its service life. So, although it participates fully, from a material point of view, in each production process, it does not do so as value. Only a portion of its value is transferred; hence, only a portion enters circulation. Thus, only this portion is circulating capital. The remaining value that has not yet entered circulation is fixed capital.

Once this aspect is clear, the connection with the category of turnover can be established. The turnover of capital is only fulfilled when the entire capital has circulated. The following example illustrates this comprehensive movement applying the categories presented. This passage can be challenging. Table 6.1 provides for any necessary visual support.

Table 6.1 Total capital turnover and circulating capital turnover

Production cycle	*Constant capital (fixed–circulating)*	*Variable capital*	*Surplus value*	*Circulated capital (c + v + s)*
I	9–1	0.5	0.5	2
II	8–1	0.5	0.5	+ 2 = 4
III	7–1	0.5	0.5	+ 2 = 6
IV	6–1	0.5	0.5	+ 2 = 8
V	5–1	0.5	0.5	+ 2 = 10
VI	4–1	0.5	0.5	+ 2 = 12
VII	3–1	0.5	0.5	+ 2 = 14
VIII	2–1	0.5	0.5	+ 2 = 16
IX	1–1	0.5	0.5	+ 2 = 18
X	0–1	0.5	0.5	+ 2 = 20

Note: Total capital = 10.5 units.

This example works with the following assumptions: the total capital is 10.5 units; of this total capital, 10 units correspond to constant capital and 0.5 units correspond to variable capital. At each production cycle, 1 unit of constant capital is released as circulating capital and the rest remains as fixed capital. Therefore, constant capital will only be fully circulated at the end of ten production cycles. It is also implied that labour-power creates enough value to restore its own value and add a surplus value of 0.5 units (i.e., the rate of surplus value here is 100%).

This picture allows us to visualise that, in each production process, part of the constant capital is released as circulating capital and part remains as fixed capital. The circulating capital will have its own turnover time: the *turnover time of circulating capital*. As one can see in the table, circulating capital turns over several times until total capital turns over just once. What determines the turnover time of the *total* capital value is the rate at which constant capital transfers its own value to the final product. It is, therefore, the service life of those elements of fixed capital.

Furthermore, the greater the initial proportion of fixed capital in relation to constant circulating capital, the longer the turnover time of the total capital value. In the example above, the constant capital of 10 units is, in the first cycle, 9 units fixed and 1 unit circulating. As a result, it takes ten production cycles to complete a full turnover. If the initial proportion were higher, for example, 9.5 units fixed and only 0.5 units circulating, the total turnover time would be 20 cycles instead of 10. Similarly, if the initial proportion were lower, say 8 units of fixed capital and 2 units of circulating capital, the total turnover time would be five cycles instead of ten. This has decisive implications for the dynamics of consumption and for understanding the ecological impact of capitalist production.

Before moving forward, there is a third category of turnover to be added to this consideration. The *average turnover time* is the time needed for the circulated capital to reach a magnitude equivalent to the initial capital.[5] Looking back to Table 6.1, the line referring to the sixth production cycle is highlighted. In the last column, the accumulated circulated value at the end of this sixth cycle is 12 units. This means that the initial capital of 10.5 units, although not yet fully circulated, has already circulated 12 units, which exceeds its initial value. In the case of our example, therefore, the average turnover time would be six cycles, since at the end of the fifth period only 10 units had circulated. It is important to emphasise this point: with each round of production, 0.5 units of surplus value are being added to the circulated value. That is why the circulated sum reaches and exceeds the initial sum even before that initial sum has completely entered circulation.

From here, this stream of thought must once again follow two parallel paths. First, the tendency to accelerate the turnover of circulating capital. Second, the tendency to prolong the production of certain commodities over time.

The category of average turnover time reveals that circulated capital reaches a magnitude equivalent to the total capital initially advanced before the total turnover time is completed. In this time period, therefore, the capital acts in the creation of value *as if it* had performed a complete turnover. This is possible because, as mentioned before, circulating capital turns over several times. If this is the case, the shorter the turnover time of circulating capital, the more times it turns over each period and, therefore, the shorter the average turnover time. And the shorter the average turnover time, the faster the amount equivalent to the capital advanced will be recovered.

To effectively introduce a variation to this process, it is worth recapitulating Table 6.1. In its presentation, it was implicit that circulating capital only turns over once per period. Therefore, for example, if the period were one year, the turnover time of circulating capital would be one year. However, it does not have to be this way. There could easily be a shorter turnover time, say ½ period (or six months). Under these conditions, the turnover times would change. Table 6.2 shows that except for the change in the circulating capital turnover time, the other hypotheses remain the same.

Evidently, if the turnover of circulating capital is accelerated, average turnover time decreases and total turnover time decreases. In other words, the time in which capital returns in a magnitude equivalent to the initial one decreases, and the time in which it returns completely, and expanded, decreases. In light of this, it is obviously in the interest of capital that the turnover time of circulating capital is as short as possible. In other words, for circulating capital to circulate ever faster.

Table 6.2 Accelerated capital turnover

Production cycle	*Constant capital (fixed–circulating)*	*Variable capital*	*Surplus value*	*Circulated capital (c + v + s)*
I	9 − 1	0.5	0.5	2
	8–1	0.5	0.5	+ 2 = 4
II	7–1	0.5	0.5	+ 2 = 6
	6–1	0.5	0.5	+ 2 = 8
III	5–1	0.5	0.5	+ 2 = 10
	4–1	0.5	0.5	+ 2 = 12
IV	3–1	0.5	0.5	+ 2 = 14
	2–1	0.5	0.5	+ 2 = 16
V	1–1	0.5	0.5	+ 2 = 18
	0–1	0.5	0.5	+ 2 = 20

Note: Total capital = 10.5 units.

Put differently, capital tends to speed up the circulation of circulating capital because this speeds up the production of value and surplus value. For each period, a given capital that circulates faster will produce a greater mass of value than another of the same size that does not circulate as fast. Marx calls this relationship between the frequency of turnovers and the surplus value created the *annual rate of surplus value.*[6] It is possible to visualise it by comparing row I of Table 6.1 to row I of Table 6.2. When circulating capital performs only one turnover per period, the circulated value is 2 units, of which 0.5 units correspond to surplus value. When circulating capital performs two turnovers per period, the value circulated at the end of the period is 4 units, with 1 unit corresponding to surplus value.

Thus, the same capital of 10.5 units may generate a greater amount of surplus simply by circulating faster. This changes the pace at which capital must carry out changes of form. As mentioned before, given that capital exists simultaneously in its various forms and given that the continuity of the repetition of cycles is necessary, the fluidity and synchronisation of the changes in forms also become necessary. The changes must occur in a coordinated manner, at all points in the cycle, and at a *determined pace*. Reducing the turnover time of circulating capital increases this rhythm. It, therefore, increases the need for fluidity in the whole process because it increases the number of times the entire production cycle is renewed. The second rotation of the period can only start once the first has been completed. The first rotation of the following period can only start once the second rotation of the previous period has been completed. And so on.

In short, due to the different circulation characteristics of each portion of capital, there is an impetus to reduce the circulation time of circulating capital. This acceleration in the turnover of circulating capital brings with it an increase in the pace of transitions in form and, therefore, an increase in the requirement for fluidity in these transitions.

To draw ecological implications from this reasoning, it is enough to remember that one of the many transformations that capital has to pass through repeatedly, and more and more frequently, happens when the commodities produced are sold. This whole movement is absolutely incompatible with moderate global consumption.

However, there is a second path to follow in the stream of thought based on the categories related to turnover. The average turnover time informs how long capital takes to return in a magnitude equivalent to the one advanced, the initial capital. However, this initial capital (the total capital value) only ends its entire turnover period when all the capital that was initially fixed completes its turnover cycle. The capital initially advanced can only return in full once it has fully entered circulation. In other words, when all the capital that was initially determined as fixed capital is extinguished *as fixed capital*. In Marx's words: "To the extent that fixed capital exists as such, [it] does not return, because it does not enter into circulation. To the extent that it does enter into circulation, it no longer exists as fixed capital."[7] In Tables 6.1 and 6.2, this corresponds to the moment when the value in the "fixed constant capital" column reaches zero.

Thus, the longer the means of production last, the longer the total turnover time. But the critical point to emphasise here is that the length of the total turnover time, which is determined by the turnover time of fixed capital, determines *the length of time during which circulating capital is tied to a specific material form.*

Because of its decisive ecological implications, this conclusion deserves further elaboration. Circulating capital reconverted into money at the end of its own turnover period could theoretically be applied in any other branch of production, taking on another material form and participating in the production of different things. However, the necessary continuity and fluidity of the cycle of capital as value also require the same kind of continuity in material terms. At least until all the fixed capital has circulated. Therefore, the service life of the means of production defines, to a large extent, the minimum time that capital is tied to the production of specific commodities. But this reasoning may be presented in a more straightforward way.

The longer the turnover time of fixed capital, the greater the number of turnovers that circulating capital will have to perform until the total turnover time is completed. And not only will circulating capital have to make a greater number of turnovers, but it will have to do so by functioning materially in the same way, i.e., *by participating in the production of the same commodities.*

Here, it is necessary to precisely define what temporal requirement means. From the fact that capital must remain tied to a given material form for a minimum amount of time, producing commodities of the same type unfolds the requirement to systematically renew the need to consume that commodity over a minimum time period. This minimum time will be longer the longer fixed capital remains in this sphere of production.

This creates substantive obstacles to the elimination of certain forms of consumption. For capital, it is intolerable that certain commodities are no longer consumed, not only from a quantitative point of view but also from a qualitative point of view. For example, the end of the passenger car is intolerable not just because the owners and managers of the big car manufacturers are environmentally unaware and greedy, or because governments are negligent, or because there is little popular pressure. All this is true in most cases. However, the end of the passenger car is intolerable because it would involve an extraordinary destruction of fixed capital. And this is not about the material existence of this fixed capital, such as buildings and machines. What is meant here is capital, in its strictest sense, as self-valorising value.

To conclude, another relevant example of the importance of this time determination is the entire energy sector infrastructure. The service time of the production, transmission, and distribution apparatus related to energy and electricity is usually more than 50 years. By looking at all the installed capacity related to the fossil economy, it becomes clear that the goal of the energy transition faces a titanic enemy. Considering that this installed capacity continues to grow vigorously, the scenario becomes even bleaker.

In the last section, when discussing the relationship between the speed of turnover of circulating capital and the creation of surplus value, an important category was briefly mentioned: the annual rate of surplus value. This happened during the analysis of the fact that the circulation of circulating capital tends to accelerate because this drives the creation of value and surplus value. In other words, for each period, a given capital that circulates faster will produce a greater mass of value than another, of the same size, that does not circulate as fast.

There is now a new conclusion to be drawn from this same reasoning. The magnitude of the capital advanced needed to produce *a given mass* of surplus value during a given period becomes *smaller* as the number of turnovers increases. This is a different insight. Before, it was assumed that the capital advanced was given and concluded that the creation of surplus value would be greater if this capital circulated faster. Now, the assumption is that surplus value is given and the conclusion is that the faster capital circulates, the smaller its magnitude has to be to achieve that surplus value. According to Marx: "by turning over ten times, and hence repeating its advance ten times, the capital of £500 performs the function of a capital ten times as great, a capital of £5,000."[8]

The total mass of surplus value produced is then determined by a new factor: the number of turnovers performed in a given period. As turnover time decreases, the number of turnovers performed increases and, with it, the annual rate of surplus value. Consequently, the mass of surplus value produced in each period increases.

As previously mentioned, turnover time is the sum of production time and circulation time. A small parenthesis might be necessary. This circulation discussed is the circulation of commodities. In other words, it is the moment when capital is in the form of money, seeking to acquire means of production and labour-power, or the moment when capital is in commodity form, seeking to realise the sale. When, in other contexts, the circulation *of capital* is discussed, that refers to the passage of capital through all its forms, including those acting at the moment of production.

If reducing turnover time boosts accumulation, shortening the periods in which capital remains in the circulation of commodities also works in this direction. Such a reduction contributes to expanding the mass of value simply by making it possible to renew the cycle more frequently. In other words, the frequency with which the production and launching of commodities onto the market can be successfully renewed depends to a large extent on how long capital spends in commodity circulation.

Similarly, reducing production time also speeds up turnover, increasing the frequency with which circulating capital needs to be advanced. This is usually achieved by reducing labour time, increasing the efficiency of the productive apparatus, and increasing the productivity and intensity of labour. In short, by raising the productive forces.

As demonstrated in Chapter 4, an increase in the general productivity of the production process increases the mass of commodities produced in relation to the magnitude of the capital employed. Considering that capital encounters an obstacle in the consumption capacity at a given time and that a capital that turns over faster delivers the same result in a shorter period as a larger capital that turns over more slowly, an important implication is drawn: for each level of society's consumption capacity at a given time, less and less capital will be employed.

In other words, since the market is only capable of absorbing a given volume of commodities at each time, and since there is a tendency for turnover to accelerate, the amount of capital capable of producing this volume of commodities is ever smaller. In this respect, Marx says: "because of a doubling of productivity, a capital of only 50 needs to be invested where 100 was needed before, and a capital of 50 and the necessary labour corresponding to it are released."[9]

In other words, increased productivity liberates not only labour-power, but also capital. This liberated capital cannot settle into immobility. It needs to continue its expansion movement, without which its very existence is annihilated. To do this, it must seek out and develop other branches of production that are not yet saturated. Or, more importantly, create entirely new branches.

Consequently, the *sphere of* circulation must be expanded in *three ways* (including the dimension related to scale). First, quantitatively, by expanding consumption without significant changes in the variety of commodities consumed. Second, also quantitatively, by expanding consumption through the dissemination of needs that were previously restricted to layers of society with greater purchasing power, i.e., the spread of consumption patterns that already existed but were previously enclosed in a restricted circle of consumption. Third, qualitatively, by creating, developing, and stimulating completely new needs.

The first way manifests itself as increasing prodigality within certain consumption habits. This is basically consuming more of the same. Several well-known sectors may illustrate the second way. Take the car industry, for example. Initially, they were totally inaccessible to the majority of the population. Then, for decades, the cars considered popular were models without air conditioning, power steering, electric windows, etc. Today, these items are progressively being included as basic items, thus becoming "popular." Other clear examples are the fashion industry, the household appliances and electronics sector, the tourism industry, etc. The third way is undoubtedly the easiest to exemplify because virtually every day, we are flooded with urgent needs that, until yesterday, we did not even know existed.

In conclusion, capital has an imperative of valorisation on an expanded scale that implies a tendency towards continuous expansion in the scope of production. Also, this production expanded in scope needs to be realised on the market. Therefore, the survival of capital depends on the continuous realisation of this expanded scope production on an expanded scope consumption sphere. In the same sense, Mészáros highlights, as a result of these same needs, the tendency towards the normalisation of luxury, the expansion of the sphere of consumption towards increasing prodigality standards.[10]

The last of the points to be dealt with in this discussion of consumption is the requirement related to speed. This goes back to the initial discussion of reducing turnover time by shortening circulation time. From the fact that capital exists in all its forms simultaneously comes the conclusion that reducing the time it spends in one of its forms means, at least momentarily, increasing the *relative* time it spends in the other forms.

In other words, if the time capital spends in the circulation of commodities is reduced, the time it spends in the production process, the moment of value creation, is *relatively* increased. In other words, fractions of capital return to the form of productive capital more often.

However, to start a new period of turnover, capital must always go through the entire cycle comprising production time and circulation time. Therefore, the more successful the reduction in the time capital spends in commodity circulation, the greater the number of times capital passes through the productive sphere and, more importantly at this point, the greater the volume of commodities produced in each period. It is not hard to see that this increases the pressure on the market to the same extent.

In other words, sales needs increase at the same time as they are (and *because* they are) satisfied. Because capital now turns over faster and spends relatively more time on production. So, on the one hand, the expansion of production requires the expansion of the sphere of consumption; and, on the other hand, the increase in the number of turnovers requires an increase in the speed of consumption, i.e., the renewal of the need for new consumption.

Mészáros has a well-known reflection on this subject. According to him, societies in general (not only the capitalist ones) always have to allocate labour and available resources to produce non-durable goods and to produce durable goods. In the course of the development of the productive forces, says the author, the allocation tends to favour the production of durable goods. However, for the series of reasons already discussed in this topic on consumption, this tendency towards durability directly challenges the self-expanding imperatives of capital.

The mode of production governed by capital, says Mészáros, is the enemy of durability. It acts (and needs to act) actively and deliberately to erode durability and discourage practices geared towards durability and reuse. Under no circumstances can the market become saturated by goods remaining in the sphere of consumption for too long.

In line with this logic, commodities must take on ever more ephemeral forms. Capital achieves this in various ways. At the current stage of its development, some stand out. The best known is planned obsolescence, which consists of deliberately reducing the service life of the goods produced. There is also the possibility of directing resources towards the production of goods that, due to their material characteristics, exclude (or to a large extent reduce) the opportunities for shared consumption, for example, encouraging the development of the passenger car industry as opposed to encouraging large public transport systems. There is also the (albeit relative) increase in the cost of maintenance and repair practices. Finally, there is something called moral obsolescence, which differs from planned obsolescence as it does not necessarily plan a shorter service life for the commodity during its production. It operates via advertising and continuous launches of new models and updates. In other words, it achieves obsolescence in the subjective sphere, influencing the perceived needs of the consumer.

The discussion on technology shows that the rationalisation of consumption in the individual production process is a necessity. In other words, capital, in its insatiable thirst for surplus value, needs to save means of production by increasing efficiency, it needs to save living labour by increasing productivity, and it needs to save time by increasing the intensity of labour. However, this imperative to save on the level of the individual capitalist is countered by the wasteful tendency of the system as a whole. At the level of the whole, what we see is an incessant expansion of production, consumption, wasteful consumption, and premature disposal of all kinds of commodities. And all of this is also a necessary result of the logic of the system and, more specifically, of the needs arising from the turnover of capital.

This is why Mészáros states that the contemporary capitalist society finds "the equilibrium between production and consumption necessary for its continued reproduction only if it can artificially 'use-up' at a great speed (i.e. prematurely discard) vast quantities of commodities which formerly belonged to the category of relatively durable products."[11]

This brings down the uncritical defence of economic growth combined with the so-called conscious consumption, which, in general terms, proposes reducing the quantities consumed and increasing the use rates of objects, increasing their durability. As long as capitalist production predominates, these expressions of individual ecological awareness will always be subordinate, however virtuous they may be. In other words, they will always be run over by the growth engine of capital. In conclusion, the so-called conscious consumption will only become possible on a global scale if we overcome capital.

Notes

1 Marx & Engels, *Collected works*, vol. 5 (1976).
2 For the same discussion, in greater detail, see Sá Barreto, "As múltiplas dimensões da exigência imposta à esfera do consumo pela dinâmica do capital: elementos para o debate ambiental" ["The multiple dimensions of the requirements imposed on the sphere of consumption by the dynamics of capital: elements for the environmental debate"] (2014).
3 Marx, *Capital*, vol. 2 (1992).
4 Ibid.
5 Ibid.
6 Ibid.
7 Marx & Engels, *Collected works*, vol. 29 (1987), 106.
8 Marx, *Capital*, vol. 2 (1992), 383.
9 Marx & Engels, *Collected works*, vol. 28 (1986), 335–336.
10 Mészáros, *Beyond capital* (1995).
11 Ibid., 550.

References

Marx, K. (1992). *Capital* (Vol. 2). London: Penguin Books.
Marx, K., & Engels, F. (1976). *Collected works* (Vol. 5). New York: International Publishers.
Marx, K., & Engels, F. (1986). *Collected works* (Vol. 28). New York: International Publishers.
Marx, K., & Engels, F. (1987). *Collected works* (Vol. 29). New York: International Publishers.
Mészáros, I. (1995). *Beyond capital.* London: Merlin Press.
Sá Barreto, E. (2014). As múltiplas dimensões da exigência imposta à esfera do consumo pela dinâmica do capital: elementos para o debate ambiental. *Revista da Sociedade Brasileira de Economia Política, 37*, 143–168.

7 Market mechanisms (to solve market maladies)

This chapter will revisit some elements that have already been discussed a few times, especially throughout Chapter 1. It presented the way economic thinking on the environment, despite its internal variety, develops along at least four well-defined axes. Chapters 3–6 focused on the issues of growth, technology, and consumption. The last of these four axes is the faith in market mechanisms not only as optimal regulators of economic life but also as optimal regulators of the metabolism between humanity and nature.

The structure of this type of reasoning is quite simple. In it, there is a basic diagnosis and a basic prescription. From the point of view of the diagnosis, the environmental problems we face are seen as the result of either a market failure or the simple non-existence of a market. From the point of view of the prescription, these environmental problems would be solved by correcting any market failures or by creating a market from scratch.

Therefore, the basic diagnosis is split into two and the basic prescription is split into two as well. Each of these branches will drive different reflections, especially from Environmental Economics. After briefly commenting on the idea of market failures and their corrections, this chapter will focus on the idea of the absence of a market and its creation.

The concept of market failure is tattooed on economists' heads throughout their education. But as this book aims to reach a wider audience, it is worth exploring the concept, taking the opportunity to connect it to the discussion. It is not a widely known fact that, for the standard economist, the market, when functioning ideally, would be able to spontaneously and automatically perform the most efficient distribution of resources. The concept of efficiency implied here is economic, not ecological, which will be discussed later in this chapter. For now, however, this distinction is not necessary. Market failure occurs when, for some reason outside the normal functioning of the market, the most efficient allocation is not achieved.

From an ecological point of view, market failure usually translates into some form of environmental damage. In this reading, environmental damage occurs

DOI: 10.4324/9781003512516-8

because the agents involved are not confronted with the environmental costs of their activities. In other words, private agents extract economic benefits from their activities and bear the economic costs of these activities. However, their activities also have environmental costs that are not borne by them as individuals but by the community as a whole. These costs are, therefore, not included in the rational economic calculation. As a result, economic rationality would end up not acting to generate the most efficient result possible. Since this non-efficient result is the result of not incorporating all the variables into the rational calculation, economists call it an externality.

If the problem rests in the fact that important information is outside the calculation, then the solution could not be any other: find a way to bring this information into the calculation. Economists will call this process the internalisation of externalities. The expectation is that once all the costs of a given activity are considered, that activity will operate at a different level (or in a different form) to the one that generates the environmental damage. And with this, an ecologically sustainable result would be achieved.

Subsequently, a question arises: how can agents be forced to incorporate considerations they ignore under normal conditions? Many different answers will be given to this question, but three main lines stand out. There is the Pigouvian tax, named after the economist Arthur Cecil Pigou, which is based on the polluter pays principle, the cost–benefit analysis, and the distribution of tradable pollution permits. The latter is also one of the bases of the second part of the chapter, which discusses it on a deeper level.

The polluter pays principle advocates that if a tax were imposed on the generator of pollution corresponding to the level of pollution generated, it would push this agent to generate less polluting waste. Either by reducing their level of activity, by reconfiguring the way they produce, or by treating their effluents to reduce their polluting potential. This type of reasoning originates in the more orthodox theoretical tradition of economics. Sadly, it is also defended by critical thinkers. To cite just one emblematic example, Michael Löwy advocates a carbon tax as a tactic to buy time until society can finally move forward.[1]

However seductive this idea may be (especially when committing to it only from a tactical point of view), the theoretical elements provided in this book are already enough to dismantle it easily. The discussion about efficiency in the chapter on technology exposed that higher efficiency saves resources, but it also saves capital. Consequently, because of the expansion-oriented nature of capital, this released amount needs to find other spaces to carry out its own logic, which ultimately neutralises the initial saving effect. It is easy to see that the carbon tax and any other tax that follows the same reasoning offer an analogous case.

It is claimed that the tax would force economic agents to redo their calculations. Once this happened, the material demand from that capital would somehow be reduced (even if only proportionally). For the purpose of analysis, the result of reduced material impact is here considered. Without it, the argument favouring the tax would not make sense from the outset. So, once the material demand for capital is reduced, the same process of releasing capital would occur. Advances that used to be made now would no longer have to be made. The sequence is exactly the same. This released capital, prevented (or discouraged) from operating in that activity because of the new tax levied, would have to find another space to operate. In doing so, it would re-establish contact with materiality.

Of course, since many types of pollution have an imminently local impact, a tax like this can successfully reduce, or even solve, a *local* pollution problem. However, it is undeniable that, on the one hand, this solution merely displaces the impact to a different location and that, on the other hand, it does not apply to global problems (not even in this very narrow sense). It is therefore anachronistic to embrace this kind of "way out," not only from a theoretical point of view, but also from a practical one.

On the subject of cost–benefit analysis, there is a quote from one of its main formulators which says enough. It does not even need to be criticised at length. This is what Ronald Coase says in his classic 1960 text, "The Problem of Social Cost":

> If we assume that the harmful effect of the pollution is that it kills the fish, the question to be decided is: is the value of the fish lost greater or less than the value of the product which the contamination of the stream makes possible.[2]

In other words, environmental damage is not evaluated in itself, in terms of its qualitative effects. It is evaluated against the economic benefits of the activity that causes environmental damage. If the economic benefits are great enough, who cares about the fish?

The third type of response to the need to internalise externalities is obviously pertinent to the discussion of market failures. However, the distribution of tradable pollution permits is key to the discussion of the absence of a market and the need to create one. The second part of this chapter focuses on it.

The defence of tradable pollution permits is based on the scenario known as "The tragedy of the commons," which is usually associated with Garrett Hardin.[3] The logic illustrated by this scenario is that when property is shared, everyone can enjoy the benefit it brings. In this way, no one can be excluded from this usufruct. However, since the property is common, it would be plausible to assume that each individual would have very little access to its benefits since everyone has access to it and no one can be excluded. This lack of equal access would produce a tendency in individuals to increase the scale and speed of their appropriation. Thus, these individuals would secure a more substantive appropriation for themselves before their peers could make their claims. Of course, since this logic is available to everyone, the result is that everyone would act in this way, thus squandering the common property. In short, the tragedy of the commons states that when ownership is collective, there is a tendency to overexploit the resource. When something belongs to everyone, it does not belong to anyone; and if it does not belong to anyone, it is offered to those who are most ambitious and voracious in their use.

In addition to collective ownership, this story presupposes something else: the absence of a market. Since ownership is collective, the resource is simply available. Whoever arrives first makes better and broader use of it. Those who arrive later may not find anything else available. So there is a race to get there first. The result of this race is, as we have seen, overexploitation and eventual exhaustion of the resources.

And why are market mechanisms supposed to defuse this bomb? In the case of the distribution of tradable pollution permits, this common property resource would be a certain available capacity of the ecosystem to absorb waste from human activities. The logic is precisely the same as the above. There is a tendency to excessively dump waste because the receiving bodies are nobody's property. Nobody has an incentive to preserve that receiving body because nobody owns it. When proposing that tradable pollution permits be distributed, this is a distribution of property rights, which could then be traded on a market designed explicitly for this purpose. It is essential to understand its operation.

Ideally, the pollution limit would be defined according to some ecological criterion. This would provide a cap; the maximum amount of pollution allowed. Once the *cap* has been defined, each participant in activities that use that receiving body would receive a quota of that cap. Strictly speaking, this quota establishes how much of the absorption capacity of that receiving body each participant can use. In other words, each participant would now own a fraction of that capacity.

And what is to be done with this new property? Does everybody start dumping waste to the extent that they are allowed to? Actually, that is not what is expected to happen. The expectation is that some of these participants will have lower waste abatement costs. So they could, in principle, emit below their limits. But what would make them do this? According to the proponents of this reasoning, two things: first, the fact that some of the participants will have very high abatement costs and, therefore, will not succeed in respecting their own limits. Second, the existence of an arrangement in which participants with lower costs may sell part of their quotas, which was not used, for participants with higher costs. In other words, the existence of a market where these pollution permits are traded, i.e., the possibility of *trade*. That is why the arrangement described is known in literature as *cap-and-trade*.

The fragility of the argument becomes evident from a practical point of view. Its reasoning may only preserve some semblance of validity if one assumes that the initial cap is based on ecological criteria. In practice, this is not the case. The initial criteria are, as a rule, economic, and the "caps" (emphasis on quotation marks) are consolidated according to the negotiating capacity and power of the parties involved.

Leaving this problem of origin aside and assuming that the cap determined is in fact compatible with the sustainable reproduction of the receiving body, a question remains: does everything else happen according to the way it is described?

It definitely does not, but that is not so obvious. The analysis of capital provides the basis for a clear and consistent answer. Two facts substantiate this negative reply. First, the distribution of property rights does not guarantee a reduction in the polluting impetus. Second, even if we accept that the generation of polluting waste is contained as a result of this distribution, the limits thus created are irremediably subordinate to the demands of capital valorisation and therefore have no necessary relationship with environmental or ecological limits.

For the whole arrangement to work, participants need to be able to reduce the volume of waste generated, even those who will eventually need to buy extra certificates to complete their reduction efforts. There are basically three ways of reducing the discharge of waste into receiving bodies: reducing the generation of this waste, treating the waste generated, or utilising the waste for other uses.

There are two paths for the first way (reducing waste generation). The first is to reduce the scale of activities without changing the technical base. There is no need to spend much time on this option; there will never be sufficient incentives for capital to contract its scale voluntarily. In non-revolutionary conditions, this only happens in times of crisis or when a given capital is destroyed by competition. The second is an old favourite, efficiency. Again, this is a conversation that can take a shortcut.

The chapter on technology discussed how efficiency gains fulfil specific objectives of capital in its movement towards valorisation. Efficiency gains can boost the value appropriated by capital by creating a differential between the constant capital factually consumed and the value corresponding to the constant capital that reappears in the commodity. It can also be a means of slowing down the deterioration of the elements of constant capital as levels of intensity increase. And it can also be a means of increasing relative surplus value. Considering all of this in detail, the conclusion was that the same result presents itself in any of these scenarios: the material impact is not restrained. It tends to increase.

However, efficiency gains have a last refuge: waste control. After everything that has been presented, it may seem strange to admit that capital will always seek, according to its own rationality, to contain wastage. This can be demonstrated. By this point in the book, it is clear that the magnitude of a commodity's value is determined by the socially necessary labour time. Also, that this labour is not limited to living, directly employed labour. The labour objectified in the means of production is transferred to the end product, reappearing in the commodity. The magnitude of this transfer is determined by the amount of means of production consumed under the average conditions of production.

Because this part is crucial, it requires a longer path of exploration. In Chapter 5 of *Capital*, when analysing the labour process in general, Marx demonstrates that all use value produced by labour is objectified labour. This is important because not only end products are objectified labour; intermediate products, with a few exceptions, are, too. Therefore, objects and means of labour are also objectified labour.

Use values are things that have certain useful properties that enable them to fulfil certain needs. Thinking specifically about the means and objects of labour, what happens when a product whose use value serves the production process does not fulfil this function for some reason? According to Marx, it becomes subject to the destructive forces of the natural metabolism. In other words, through the action of degradation processes (more or less rapid), the product is destroyed without having fulfilled its function. The labour objectified there, therefore, is also destroyed. It is not preserved. For this reason, Marx says:

> A machine which is not active in the labour process is useless. In addition, it falls prey to the destructive power of natural processes. Iron rusts; wood rots. Yarn with which we neither weave nor knit is cotton wasted. Living labour must seize on these things, awaken them from the dead, change them from merely possible into real and effective use values.[4]

When these things are used in the labour process, they are obviously consumed. But they are consumed according to a purpose, according to the proper realisation of their use value. In other words, objects and means of labour consumed this way cease to be merely potential use values and become actual/effective use values.

More importantly, they become part (to a greater or lesser extent) of a new use value. In other words, the labour that was previously objectified in these intermediate products is now objectified in a new use value. This is a fundamental notion. Past labour, objectified in the objects and means of labour, is *transferred* from the means of production to the end product by living labour.

This insight draws from a moment in Marx's analysis in which he deals with the labour process in *general*. This means that the reasoning presented here is valid for human productive activity *in general*, regardless of its social form. Accumulated past labour always reappears in the end product. In capitalism, *moreover*, this objectified labour also appears *as value*, forming the value of the commodity. This is why Marx says that the value of the means of production is transferred to the commodity.

Resuming the stream of thought, which was momentarily paused to clarify this last point, the magnitude of the transfer of value is determined by the volume of means of production consumed under the average conditions of production. This means that if there is a tremendously high consumption of means of production, and if this consumption is in line with the current normal conditions of production, then *there is no wastage*. Likewise, if there is relatively moderate consumption, and if this consumption is above what is required by the average conditions of production, then there is wastage.

Therefore, the notion of wastage is not predicated on material consumption, as it might seem at first glance. What cannot be wasted in any way is capital. This is why the material overexploitation of nature is perfectly compatible with the absence of wastage, from the point of view of capital. For this reason, the capitalist drive to reduce waste as much as possible bears no necessary relation to any idea of ecological sustainability.

To sum up, this chapter addressed the claim that the distribution of tradable pollution permits is an arrangement capable of curbing pollution, therefore approaching the question of the alternatives available to reduce waste generation. One of them is reducing the scale of operations, an alternative ruled out as unfeasible since it challenges the very nature of capital. The other is increasing efficiency. By analysing various possible motivations for increasing efficiency, it became clear that, in all cases, the result is not the containment of the polluting impetus.

Any technical modification that increases efficiency must be technically feasible, of course. However, this is only a necessary condition, but not a sufficient one. Any technical change that increases efficiency must also be *economically* viable to be effectively implemented. In other words, it needs to be able to provide for the expansion of capital. If the expansion of capital is a prerequisite for the implementation of more efficient techniques, and knowing the effects of this expansion, it is easy to conclude that not even in this idealised arrangement of pollution permits do efficiency gains work as promised.

The next pollution control alternative to be addressed is waste treatment. The assumption here is that waste generation remains the same and the polluting impact is contained due to some treatment given to the waste that is generated, making it less polluting. This option has the distinguished characteristic of operating at the end of the production process. Therefore, the techniques used for this purpose are known as *end-of-pipe*.

If, in the previous case, there was pressure for capital to eliminate wastage (even if only *capital* wastage), in this case, *anything* that is done counts as capital wastage.

First of all, it is important to understand waste generation in the general context of the movement of capital. In this regard, Marx says the following:

> Suppose that in spinning cotton, the waste for every 115 lb. used amounts to 15 lb., which is converted, not into yarn, but into "devil's dust". Now, although

> this amount of waste is normal and inevitable under average conditions of spinning, the value of the 15 lb. of cotton is just as surely transferred to the value of the yarn as is the value of the 100 lb. that form the substance of the yarn. The use value of 15 lb. of cotton must vanish into dust before 100 lb. of yarn can be made. The destruction of this cotton is therefore a necessary condition for the production of the yarn. And because it is a necessary condition, and for no other reason, the value of that cotton is transferred to the product. The same holds good for every kind of refuse resulting from a labour process, where that refuse cannot be further employed as a means in the production of new and independent use values.[5]

Once again, it is clear that what is commonly understood as material wastage does not necessarily count as capital wastage. This passage, however, seems to pose an apparent contradiction with the argument that has been built up so far. In the chapter on technology (and in this one, too), one of the things discussed was how the gain in efficiency provides an advantage for capital because by using fewer raw and auxiliary materials than average, there would be a difference in value between the component of constant capital that reappears in the commodity and the constant capital actually employed. The practical implication of this is that there is an incentive for each capital to seek to operate at a level of efficiency that is always above average, which implies seeking to reduce the consumption of raw materials and auxiliary materials as much as possible.

As demonstrated, this does not have the material effect expected by optimists because reductions like this are only economically viable if they increase capital. But if this impulse to reduce the consumption of raw and auxiliary materials as much as possible is based on the logic of value transfer, and if the value of these materials that become waste also reappears in the commodity, then wouldn't there be the same kind of impulse to reduce waste?

Efficiency gains do have the effect of *proportionally* reducing waste *generation*, which has been previously discussed. The question now concerns waste reduction via *end-of-pipe* techniques. The logic of appropriation is key to properly understanding this reference. The presentation structure employed by Marx in *Capital* allows for a brief and sufficient discussion on the matter.

The argument developed throughout the three volumes of *Capital* is built at different levels of abstraction. Within each of the books, the reader also finds different levels of abstraction, but generally speaking, one may say that Book I presents the highest level of abstraction, Book II presents an intermediate level, and Book III reaches a level of abstraction that brings the reader considerably closer to everyday categories. This means that, especially in Books I and II, the discussion involves categories that make up the essence of capitalist reality, on the one hand, and momentarily disregards some categories and tendencies, on the other. Therefore, there will be debates to which the categories in Books I and

II will be useful but insufficient. The discussion of consumption was an example and something similar will happen as appropriation is discussed.

The dispute between capitals over the appropriation of socially produced wealth involves very concrete categories such as price, profit, and rate of profit. The categories of value, surplus value, and others form the essential layer for understanding the mechanics of the movement of this society. Still, they cannot fully explain how everyday practices produce this movement. The reason is that people are very rarely aware of these more abstract categories. As such, they are not the ones to spur them into action. The capitalist acts in the world with profit in mind, not surplus value. Workers act in the world with wages in mind, not the value of their labour-power.

How does Marx go from one layer to the other? It is quite a long journey that takes three books, but something more concise will be enough here and can be found in a well-known aspect of his presentation structure. In Chapter 4 of Book I, Marx tries to demonstrate the origin of surplus value. The first step of his argument is to assume that someone appropriates surplus value by either selling the commodity above its value or buying it below its value. After a few exercises, he concludes that this unequal exchange cannot explain surplus value since there is no creation of new wealth, only redistribution of pre-existing wealth. If unequal exchange does not explain surplus value, it becomes necessary to filter it out, i.e., to disregard it. In other words, the explanation of surplus value must be possible assuming the exchange of equivalents. That is, the explanation of surplus value must be possible assuming that commodities are sold for their values. Another way of describing this moment of reflection is to say that as long as Marx is still investigating the dynamics of wealth creation, he may move forward through this simplification of the exchange of equivalents. However, at a certain point, it becomes clearly insufficient. This happens precisely at the moment of approaching the dynamics of appropriation.

To arrive at the intermediate category of price of production, Marx presents us with the categories of cost price (which is the undifferentiated sum of expenditure on constant and variable capital) and rate of profit. At a certain point in the presentation of these categories, one result becomes clear: by following the level of abstraction maintained until then, one would be forced to conclude that capitals of equal magnitudes, but with different compositions of constant and variable capital, would have different rates of profit. As Marx himself warns, this runs counter to observable reality. In practice, there would be no such variety of profit rates. Quite the opposite.

Were Marx a positivist, he would have had to admit that his thinking had come up against an internal inconsistency, which required an effort to reformulate it. However, Marx was not a positivist and was, therefore, in a position to realise that the contradiction was presented by *reality*, not his thinking. And in this case, the explanation is simple: the logic that governs the *appropriation of* wealth does not coincide with the same tendencies that explain the *creation of* that wealth. In other words, a particular capital does not exactly appropriate the surplus value created under its command. It appropriates the existing surplus value produced by all the different private capitals according to an average rate, which tends to be established through competition.

Therefore, there is no qualitative difference between constant and variable capital in the capitalist's mind, which is fundamental to explaining how past wealth is conserved and new wealth is created. In the capitalist's mind, we only find the undifferentiated sum of capital that needs to be advanced and the current normal rate that can be applied to this sum. However, part of our previous discussions is of use here. The anchor of this whole dynamic is still the average conditions of production. In other words, it would not make sense to make extravagant advances of capital with a view to higher profit. In this case, advances that exceed the average conditions of production would simply be superfluous capital expenditure. They would not legitimise any additional claim on socially produced wealth.

This brings us to the point that connects to the theme of this chapter. According to Sá Barreto:

> [Waste] is not additional advanced capital. It is capital which has already been advanced as raw and auxiliary materials. Any mobilisation of capital to reduce the production of waste *beyond the average* (for this sole purpose) would consist of an additional mobilisation of capital, which would therefore be greater than that required by production performed under average conditions; an increase in costs with no counterpart in terms of capital valorisation.[6]

Indeed, for any level of efficiency considered and, consequently, for any level of scrap and waste generation considered normal, there is no incentive for capital to restrict it beyond the reductions already provided by efficiency gains. The use of additional labour *specifically and exclusively* aimed at this goal would necessarily have the character of superfluously spent labour (and therefore capital). Thus, just as in the case of reducing waste *generation*, the expectation that market mechanisms will spontaneously boost waste *treatment* with end-of-pipe practices is absolutely unfounded.

Of the three possible ways to reduce pollution, two have been explored and discarded: reducing waste generation and treating the waste generated. The last one needs to be presented: reusing waste as capital. So far, this argument has been developed in a very general way, allowing a consideration of the immense variety of cases in which the waste from capital activity has polluting effects. However, for the last point of this path, concrete reality demands a more specific analysis, aimed at the defining ecological issue of our time, the climate crisis.

To simplify a topic that Chapter 11 will explore in more detail, the cause of this crisis is the chemical transformation of the atmosphere, which is mainly driven by the emission of greenhouse gases from human activities. This simple statement makes it clear that this is actually a waste problem. The greenhouse gases humanity emits are waste from human activities (not just productive activities, but mainly). Thus, everything seen so far applies to this discussion. Even the infamous carbon markets follow the theoretical architecture of tradable pollution permits, discussed in this chapter.

Moreover, gaseous waste poses specific difficulties for the point being addressed now (the reuse of waste as capital). Carbon dioxide, carbon monoxide, methane, nitrous oxide, and others are not fit for economic reuse aimed at profit. At best, they allow for some small-scale reuse, which is eclipsed by the scale of the reuse that would be necessary for there to be any positive atmospheric impact. In other words, the hope of ecological magic operated by market mechanisms once again does not stand.

The reasoning presented throughout this chapter leads to the conclusion that the *rationality compatible with the normal reproduction of* capital is perfectly able to coexist peacefully with reasonably high levels of pollution. In contrast to this conclusion, circulating among the varied range of environmentalist conceptions is the widespread idea that our disruptive and destructive interference in the planetary ecosystems results from a way of life in which *irrationality* predominates.

Employing minimal creative freedom, this way of life may be called *business as usual*, an expression that recurs in green literature. In this context, business as usual would be an irrational model of capitalism, at least on an ecological level. The problem with this kind of assessment, which captivates even relatively critical variants of environmentalism, is that it approaches the rational/irrational opposition in an abstract way, thereby conflating a *desired* ecological rationality and an *imagined* universal economic rationality.

In other words, by simply rendering the way of life irrational, the diagnosis is that ecologically destructive dynamics are a *deviation from a normal rational course*, a perversion of a possible sustainable or less devastating capitalism. The mistake here lies in not realising that *maximum economic rationality* on the level of individual capital *is expressed in a destructive way*, i.e., as maximum irrationality on the ecological level. Therefore, the relationship between rationality and irrationality should not be approached as an abnormality within capitalism but as a contradiction which is a fundamental part of its typical/normal dynamic structure.

Mészáros makes a similar argument when by stating that

> no matter how "calculating" and "rationalizing" or "economy-conscious" the particular enterprises might [...] be, in the interest of their own survival in the market place, the system as a *whole* is utterly *wasteful*; and it must continue to be so in ever-escalating proportions.[7]

This concludes the ecological critique of capitalism based on Marx, developed from Chapters 2 to 7. Next, Part II presents and discusses debates within the Marxist ecological tradition. After that, Part III approaches a few strategies for dealing with the ecological crisis.

Notes

1 Löwy, *What is ecosocialism?* (2005).
2 Coase, "Problem of social cost" (1960), 2.
3 Hardin, "Tragedy of the commons" (1968).
4 Marx, *Capital*, vol. 1 (1982), 289.
5 Marx, *Capital*, vol. 1 (1982), 313.
6 Sá Barreto, "Marx contra a fantasia 'coaseana': uma crítica ontológica ao fundamento teórico dos mercados de carbono" ["Marx against the 'Coasean' fantasy: an ontological critique of the theoretical foundation of carbon markets"] (2015), 274.
7 Mészáros, *Beyond capital* (1995), 568.

References

Coase, R. (1960) The problem of social cost. *The Journal of Law and Economics, 3*, 1–44.
Hardin, G. (1968). The tragedy of the commons. *Science, 162*, 1246–1248.
Löwy, M. (2005). What is ecosocialism? *Capitalism, Nature, Socialism, 16*(2), 15–24.
Marx, K. (1982). *Capital* (Vol. 1). London: Penguin Books.
Mészáros, I. (1995). *Beyond capital.* London: Merlin Press.
Sá Barreto, E. (2015). Marx contra a fantasia "coaseana": uma crítica ontológica ao fundamento teórico dos mercados de carbono. *Marx e o Marxismo, 3*(5), 263–278.

Part II

Debates in Marxist ecology

8 On the division of ecosocialist thought into stages

Anyone who has had any contact with ecosocialist thought has probably come across the famous division among first, second, and third stages. This structure dates back to the early 2000s and is said to have been first proposed by John Bellamy Foster. Since then, there is little doubt that this has been the predominant categorisation of the trajectory of ecosocialist thought.

It is usually presented as follows. The first stage would be characterised by the effort to graft green thinking onto Marxist thinking. The word *graft* is not used by chance. A remarkable characteristic of a graft is that it does not usually look very smooth. Thus, the very term adopted already indicates an assessment that the result of this theoretical endeavour typical of the first stage was imperfect, to say the least.

According to Foster, the main characteristic of the kind of theoretical endeavour that marks this stage would be the use of *ad hoc* formulations to bridge the gap between green and red thinking. Another sensitive issue that often emerges is that this type of procedure reveals an underlying perspective that evaluates Marx as an insufficient thinker on ecological matters or, at times, even as an anti-ecological thinker. Foster's presentation of this first stage gives the impression that Marx's contributions are only considered insofar as they allow us to bring to the discussion a dimension of class conflicts and interests, but that Marxist thought as a whole is marginalised.[1]

The second stage, on the other hand, was to overcome this mistaken understanding that Marxian thought lacked any ecological content. Precisely for this reason, what characterises the second stage is the effort to apply Marx's theoretical legacy in thinking about environmental issues. Consequently, this is no longer an attempt at grafting. If the first stage starts with Marx rejecting him, the second stage returns to Marx in order to transcend the green thinking of the time. The dialogue with green theories would be imminently critical.[2] "Would be" because, as demonstrated in the final part of this chapter, there is some debate as to whether so-called second-stage ecosocialism actually dialogues with theories and knowledge outside a certain relatively narrow circle.

DOI: 10.4324/9781003512516-10

In any case, a brief comment must be made here, momentarily ignoring the possibility that this self-description of second-stage ecosocialism is not so precise. The kind of critical engagement with existing theories in this characterisation of the second stage closely mirrors how Marx critically engaged with political economy, for example. At no point does Marx try to graft the socialist thinking that existed at the time onto classical political economy or vice versa. He indeed appropriates numerous categories presented by Smith and Ricardo, but he does not do so in a sterile way. At every moment, these categories are the object of an immanent critique, which seeks to map out the extent to which they provide an adequate understanding of reality and the extent to which they are mystified apprehensions of reality. But there is more to it. The immanent critique employed by Marx also seeks to explain why the mystifying theories are dominant. In doing so, this type of reflection always leads to a critique of a society that needs fragile (or even false) theories about itself to guarantee its own reproduction. Therefore, it is not a question of criticising theory simply to obtain better theories. Marx's theoretical critique is always a critique of this society and always reveals the need to overcome it. This was exactly the kind of exercise carried out throughout the first part of this book.

Resuming this presentation, Foster emphasises the persistent and comprehensive presence of ecological themes in Marx's work. In other texts of his prolific work, Foster goes even further and tries to maintain that there is a centrality of the ecological question in Marx's work.[3] As explained in Chapter 2, there is a certain exaggeration in the assertion of this centrality. Even so, the demonstration of the fruitfulness of Marxian thought for ecological issues does not require this kind of exaggeration.

Finally, for some time now, there has been talk of a third stage, which seeks to incorporate experiences of environmental struggles from various social movements.[4] It remains to be defined whether this can actually be seen as a third stage and whether this effort involves organically incorporating such experiences or, as in the first stage, simply grafting them on. In his presentation of the second stage, Foster claims that one of its purposes is to severely criticise the "spiritualistic, idealistic, vitalistic and moralistic emphases."[5] An interesting research agenda would be to find out how far these emphases have been eradicated from ecosocialist thinking and how far this third stage may have brought some of them back.

This presentation of the two or three stages of ecosocialist thinking is very common, but it tells very little. It is true that it describes a process of deepening and perfecting. But it still does not clarify what the most striking theoretical aspects of each stage are. It is important to extract from their texts a little more than what can be found in the concise presentations of this structure.

That is not to say that there is something wrong with concise presentations.[6] The problem is becoming satisfied with telegraphic definitions and not going beyond them at any point. By increasing the level of detail sought, many interesting aspects emerge.

Certainly, the most emblematic and remembered first-stage contribution is James O'Connor's two contradictions.[7] According to O'Connor's peculiar reading, there is a first contradiction in capitalism that Marxists widely recognise. This contradiction would lie, in a nutshell, in the impulse of individual capital to reduce costs by replacing workers with machines. In doing so, they would deprive growing fractions of the working classes of their income and thus decrease demand, creating a problem of product realisation; in other words, there would be a crisis triggered by insufficient consumption. O'Connor adds a second contradiction to this first one. According to the author, in this same drive to reduce costs, capital would seek to impose part of the costs of its operation on society as a whole. To use current economic vocabulary, O'Connor's second contradiction involves a drive to *externalise costs*. However, he claims, these externalised costs would eventually turn against capital itself, degrading and making the material conditions of production more expensive and, ultimately, reducing profits. In other words, there would be a supply crisis.

For each contradiction, O'Connor attributes a tendency towards specific forms of crisis. A hasty reading would lead to the conclusion, in line with what the author himself seems to suggest, that this second contradiction imposes a *natural limit* on the destructive march of capitalism. The exponential rise in costs would either brake or destroy capital. This is not very different from what many orthodox economists also claim. Burkett's critique of O'Connor's two contradictions is spot on and pinpoints exactly where the problem lies.[8]

According to Burkett, O'Connor's weakness is the same kind of reductionism discussed in the chapter on market mechanisms. He conflates economic limits and ecological limits. In other words, he fails to realise that when capital faces an economic limit, it may have already crossed several ecological limits a long time ago. In Burkett's words:

> Unfortunately, O'Connor's model does not distinguish the natural conditions required for profitable capitalist production from those needed for an ecologically sustainable production that can support human development. [...] As a result, the model reduces the question of environmental crisis to that of economic crisis and bypasses capitalism's ability to reproduce itself on the basis of profitable environmental maintenance activities that do not reverse the degradation of nature from a coevolutionary human developmental perspective.[9]

It is easy to demonstrate that capital does not need to guarantee the integrity of ecological conditions to guarantee its own profitable existence. As seen at the beginning of the book, the greatest possible appropriation of surplus value is the guiding purpose of capitalist production. Also, this surplus is limited, in part, by necessary labour, that is, by the time of day that needs to be dedicated to restoring the value of labour-power. A fundamental determinant of the necessary labour time is the level of the productive forces. Also, the level of productive forces is partly determined by natural conditions. The role of natural conditions is particularly clear, for example, in agriculture and soil fertility conditions. High soil fertility is incredibly beneficial from the perspective of the production of use values (food, raw materials for the production of textiles, etc.). However, given this chain of relationships, one cannot simply conclude that capital will tend to favour the most sustainable possible interaction with these natural conditions. In this sense, Marx warns:

> by no means follows, inversely, that the most fertile soil is the most fitted for the growth of the capitalist mode of production. The latter presupposes the domination of man over nature. Where nature is too prodigal with her gifts, [...] Man's own development is not in that case a nature-imposed necessity The mother country of capital is not [...] the absolute fertility of the soil but its degree of differentiation, the variety of its natural products, which forms the natural basis for the social division of labour, and which, by changes in the natural surroundings, spurs man on to the multiplication of his needs, his capacities, and the instruments and modes of his labour. It is the necessity of bringing a natural force under the control of society, of economizing on its energy, of appropriating or subduing it on a large scale by the work of the human hand, that plays the most decisive role in the history of industry [...] Favourable natural conditions can provide in themselves only the possibility, never the reality of surplus labour, nor, accordingly, the reality of surplus-value and a surplus product.[10]

The relationship between capitalist production and the ecological crisis of the conditions of human development can be understood in a similar way to Marx's analysis in the passage above. Following the same line of reasoning, it is possible to say that the ecological crisis presents itself to capital in a bipartite way: on the one hand, as paradise, which offers a vast new set of needs to be exploited, but on the other, as hell, which to a large extent subtracts (and ultimately tends to destroy completely) the ineliminable material basis of any form of social organisation.

The limit of O'Connor's analysis, according to Burkett, lies in the fact that he does not adequately understand the difference between the environmental crisis of capitalism and the ecological crisis of the material conditions of human development. Capital has many alternatives at its disposal to deal with its environmental crises, but none of them are capable of dealing with the persistent erosion of natural conditions that is pushing the system towards a real ecological collapse.

Despite Burkett's correct observation that environmental management can be profitable for capital, it is important not to lose sight of the fact that this type of management does not lead to any kind of ecological sustainability. The sustainability that interests capital, even if it engages in environmental management, is *economic* sustainability. And, as extensively discussed in the previous chapters, in capitalism, economic sustainability acts in every way as an accelerator of destruction.

The conscious distinction between environmental, economic crises and ecological crises affecting the material base of society is one of the hallmarks of the second stage of ecosocialist thinking, as it is presented. Having gone through O'Connor's contribution, albeit very briefly, it becomes clear why the so-called first stage faces chronic difficulties in realising the qualitative differences among these crises. By uncritically incorporating pre-existing green theories, the first stage smuggles a series of Trojan horses into red thinking. In other words, it admits theories that do not see (forward or backward) beyond capitalism. Thus, for example, the notion of externalities (and the internalisation of externalities) is smuggled in; the notion that nature is merely a static repository of natural resources and a sink for our waste finds its way in. There is also the incorporation of *trade-off*, cost-benefit, valuation, and new national accounts analyses.

This idea of a new accounting system, which proposes a green GDP, deserves some attention. Traditional metrics on economic performance (such as GDP) neglect the environmental dimension because *the very objective movement of this society neglects everything outside the realm of value and its movement towards valorisation.* The metric exists as it does because the reality being measured is as it is, not vice versa. In other words, GDP is a predominant measure because it most closely reflects the economistic and anti-ecological character of the society in which we live. Capital is not as destructive as it is because we use measures like GDP. Therefore, however well-intentioned and rigorously formulated, alternative metrics will not alter the dynamics of capital and the fundamental social relations that correspond to it. This is possibly the most naïve and idealistic way of proposing to internalise externalities.

Burkett, in a way, seems to embrace this idea, although eventually stating that greening the GDP is as feasible as greening capitalism itself. He says:

> From an ecosocialist perspective, the key distinctions [...] are between measured (market-valued) and real (need-satisfying) economic growth and between accumulation and legitimation crises. Both distinctions clearly imply that the environmental conditions required by capital accumulation differ from those needed for sustainable human development.[11]

The second difference is extremely important. But in the first part, there is a mistake in thinking that market value does not measure something real. This error is based on not recognising the duality of wealth in capitalism and on the assumption that wealth is simply used for value, which is not true (*in capitalism*). In capitalism, almost everything is produced as a commodity. And the commodity, as we know, is a unit of use value *and value*. As such, everything that is socially produced is not just material wealth, i.e., useful things that fulfil needs. It is also wealth in its social form, i.e., value. A capitalist society is a society in which capital is the central structuring category of the dynamics of wealth production and circulation. And since capital is *value in a self-expanding movement*, the real wealth that really matters in this society is wealth in the form of value. Measuring this wealth by market values is, therefore, not a measure that loses sight of real wealth, as Burkett suggests. On the contrary, it is *the measure* compatible with the character that wealth *actually* takes on in this society.

Another way of pointing out the same problem is to say that, both in interventions considered to be first-stage and in interventions considered to be second-stage, there is a difficulty in making a proper distinction between what is productive from the point of view of meeting different needs and what is productive from the point of view of capital. Marx himself dealt with this difference with great clarity and awareness. Indeed, in Chapters 10–13 of *Capital*, Marx demonstrates in detail how capital adapts the material forms of production to its logic. Through this process, the very character of productive labour changes, taking on specific determinations imposed by this blind logic.

Chapter 5 analysed the labour process in general, overlooking any historical specificities. It discussed the teleological character of human activity, which is to say it is directed towards (and conditioned by) a purpose. In this sense, activities will be considered "productive" if they result in use value, if they fulfil this purpose. However, in the context of the specifically capitalist production process, the mere creation of a use value capable of meeting needs is not enough. Only labour that results in the creation of surplus value will be productive. In other words, despite the best and highest aspirations, in this society, "productive" does not apply to production that only creates use values; it must create commodities. Furthermore, it is not just production that creates value, but production that creates surplus value.[12]

This is also what makes the old postulate that workers are more oriented towards use value and would, therefore, be more willing to make sacrifices to obtain environmental improvements so fragile. Burkett reinforces this idea in his text, and he is far from being the only one. The idea that workers have some natural and spontaneous vocation to resist the destructive march of capital simply does not stand. The following quote from Burkett's text provides an example:

> [workers are more likely than capitalists to be] willing to sacrifice a certain amount of commodity consumption in return for a sufficiently great improvement in environmental quality [...]. The implication [...] is that a business coalition organized to maximize aggregate profits and a coalition of workers

> organized to maximize their collective welfare would differ politically over the desirable degree of strictness of environmental standards. Capitalists would prefer a higher level of pollution than that preferred by workers.[13]

There are at least two unfounded assumptions here. First, workers would be willing to sacrifice a certain amount of commodities, and second, this is all they would be sacrificing. These two errors do not allow the author to realise that it is very plausible that workers are *not* willing (because they simply cannot be) to sacrifice anything. A recent emblematic example is the position taken by organised coal and *fracking* workers in the 2020 United States presidential elections. It has also been difficult for the Brazilian left to defend the ban on oil exploration. These examples are enough to realise that simply recognising the importance of class struggles for the real unfolding of conflicts in the ecological sphere is insufficient. There are struggles and struggles. And not every victory for the working class necessarily takes a sustainable direction. This issue will be discussed in detail in Chapter 12.

A critical discussion of the periodisation of ecosocialist thought is presented by White, Gareau, and Rudy in a paper published in 2017 by *Capitalism, Nature, Socialism,* entitled "Ecosocialisms, past, present and future: From the Metabolic Rift to a Reconstructive, Dynamic and Hybrid Ecosocialism." The authors begin by recognising the importance and impact of the debate on metabolism and metabolic rift triggered by Foster.[14] At the same time, they argue that the undeniable critical weight of this theoretical reexamination has also thrown many ecosocialist thinkers into a somewhat sterile effort to demarcate which ecological thinking succeeds in carrying on the Marxian and Marxist theoretical legacy. In other words, White *et al.* point out an exaggerated preoccupation with this kind of demarcation of well-defined stages in the ecosocialist tradition.

The authors present three points of initial critique, which are then complemented by other important comments. The first point of critique clarifies why the authors link the demarcation attempts to the discussion of metabolism. According to them, the ruler used to distinguish the first from the second stage is too narrow. They claim that everything before the debate on metabolism is relegated to the first stage or to a pre-figurative stage deemed even less developed. By this same criterion, among the most recent interventions, everything that incorporates the discussion of metabolism and metabolic rift is considered part of the second stage. Anything that deviates from this central theme is considered heretical and undesirable. Thus, interventions that are very different from each other end up grouped together (without much differentiation) in the first stage. Proceeding in this way, the periodisation into the first and second stages would simply be a mapping of convergences and divergences in relation to the tradition organised around the debate on the metabolic rift.

In the second initial point of critique found in the text, in the same sense as the previous one, the authors claim that recent interventions, also based on the Marxist theoretical legacy, end up pushed to the margins of the debate. In other words, other fruitful interpretations found in Marx beyond the metabolic rift are neglected when there is an excessive (or even exclusive) focus on metabolism. Among these other possibilities, White and his co-authors mention Marx's harsh critique of Malthus and the ideologies of scarcity, the critique of dead labour that sucks the life force out of living labour, the centrality and ubiquity of contradiction in his thinking, and the recognition that there is no untouched (or virgin) nature.

In their third initial point of critique, the authors point out that Marx and Engels established an intense dialogue with the science of their time. However, incorporating the science of that time into our thinking *today* is likely to be an anachronistic exercise. In other words, relying excessively on Marx's reading of the science of his time is problematic. A striking example of this behaviour can be found in Foster himself and his frequent reference to Liebig and his theories on soil chemistry.

As White *et al.* point out, rift scholars continue to dialogue predominantly with the science of the time of Marx and Engels. The result is a neglect of several important contributions that may be found in the so-called first stage. The authors mention the critique of capital accumulation combined with a critique of (neo-Malthusian) populism, the critique of naturalistic reductionism, the critique of ethnocentric eco-romanticism, the emphasis on class, gender, and race analyses, etc. According to them, these contributions precede both the rift scholarship and the hybrid critical tradition (which they themselves propose).

This proposal for a hybrid tradition does not receive much attention in the text discussed. Its more general content involves the defence of incorporating the most recent developments in ecological theories into considerations (as opposed to merely appropriating the mappings made by Marx and Engels). However, even in this text, there is a potentially serious problem in the rare moments when a critical hybrid tradition is defended.

Initially, they seem to propose appropriating scientific understandings from other areas. This is, in fact, an inescapable task for a critical tradition that works at the intersection of nature and society. Soon afterward, however, apparently without noticing the difference, they advocate using *multiple methodologies*. Then we would at the very least be flirting with instrumentalism, which is a type of scientific perspective completely alien to Marx. It would therefore be important to make this defence of methodological pluralism much more carefully.

Returning to the important contributions of the first stage that rift scholars have neglected, the authors go further, arguing that there is a decisive aspect in which the reflection of the first stage is superior to that of the second. According to them, regarding the first stage, there is a complete rejection of concerns with

Malthusian inclinations. The second stage would have incorporated a series of analyses that, strictly speaking, would once again give way to concerns of this kind.

This passage is particularly problematic because it is not accompanied by critical elaboration. Basically, it simply states that any discussion about limits has a Malthusian character. In this sense, a few points must be explored in more detail.

Malthus displayed a deep and deplorable anti-poor sentiment, which must be rejected without further consideration. But there were other problems with his thinking. He developed his population theory (a theory about natural/demographic limits) when the population was six times smaller than it is today. His hypothesis about the arithmetical growth of food production was wrong even for the time. Humanity's consumption of resources was *extraordinarily* low (the so-called great acceleration took place from the 1950s onwards). Thus, from the fact that the original Malthusianism was wrong and based on a vile class perspective to the fact that neo-Malthusianism deepens this error by transposing these intuitions to the present day, it does not follow that any serious analysis of limits is neo-Malthusian or wrong.

Among the various supposed symptoms of neo-Malthusianism, White *et al.* mention discussions about a *natural* load capacity. Of course, "natural" is problematic if understood as something without human mediation. However, rejecting the pure and virgin "natural" should not lead us to reject the existence of more or less narrow limits to load capacity. Among such symptoms, the authors also include the presence of discussions about *overshoot*, which is a term popularised in *Limits to Growth*.[15] It is unclear what the problem would be in addressing issues related to *overshoot*, especially after we have consolidated a reasonable scientific understanding of climate change, its critical points, and feedback mechanisms. In the same sense of listing the symptoms of Malthusianism, they also point to considerations of natural scarcity. Again, rejecting the idea of "natural" as something completely detached from what is social cannot lead to an understanding that scarcity is a *purely* social phenomenon. For example, there is the possibility of crop failures, the depletion of fish populations, droughts, etc.

It is not surprising that they superficially reject discussions about overpopulation. But why should all this be a symptom of neo-Malthusianism? Talking about overpopulation when we count the human population at almost 8 billion is qualitatively very different from talking about overpopulation when we were just over 1 billion individuals, as was the case at the time of Malthus. The authors' concern is that the discussion about limits may legitimise environmental racism and ecological imperialism. Of course, this is a possibility. However, in this case, as in so many others, it makes no sense to enter this dispute giving up on a rigorous understanding of reality. What White and his co-authors do at this point in the text is criticise interpretations by rejecting the uses that can be made of them. This is the same as, for example, criticising the theory of relativity because of the possibility of a nuclear bomb.

To conclude this chapter, it is worth recovering one last element of this text that relates to the theme of Chapter 10. The authors rightly criticise a background expectation of rift scholars that would involve re-establishing some form of balance, of original metabolism that has been lost but may be recovered. In contrast, White *et al.* argue that the rift scholarship loses sight of the fact that contemporary ecological theories no longer deal with the notion of balance, having moved towards more systemic and processual approaches. In agreement with such an assessment, one may even suggest that such a poor assessment of this kind of complexity perhaps explains the relative absence of a sense of urgency in the literature on metabolism. According to White and his co-authors:

> we end up with a form of eco-Marxism that skirts the question of how, whether or to what extent any and all local and regional socio-environmental relations and crises are, in part, the product of past rounds of socio-natural metabolisms, much less how and to what extent future ecologies are going to necessarily be profoundly shaped by current rounds of anthropogenic activity.[16]

This is crucial and, indeed, rarely addressed in the literature on the metabolic rift. It is as if, once capitalism is overcome, humans would be able to re-establish old balances. Rifts are seen as processes that can always be "fixed." They are not understood as antecedent material conditions for subsequent forms of metabolism.

Consistent with this critique of the somewhat balance-oriented approach developed by the rift scholarship, White *et al.* recognise that global warming is a *fait accompli*. Correctly, they state that

> Climate science is not simply telling us we have a crisis. It is also increasingly informing us that there can be no going back to the Holocene. [...] It is increasingly evident that in the very best scenarios, we will be living on a planet that is at least 1.5–2°C warmer.[17]

However, despite this realistic assessment of our current and future condition, the authors claim that placing the planetary collapse at the centre of political action would lead to an undesirable politics of catastrophism. This kind of reasoning, which is unfortunately not so rare, is indefensible. If there is a planetary collapse underway, and if this crisis is central to the destinies of humankind, there is nothing else that should be more at the centre of political action than it.

Evasively, the authors advocate a policy with a more idyllic imaginary instead, going as far as to mention the notion of a post-scarcity ecological society. This idea of a society that definitively overcomes the problem of scarcity strongly alludes to the notion of a supposed luxury ecological communism. These imaginaries can only be defended by assuming scarcity is a *purely social* phenomenon. This understanding, however, is patently wrong. Regardless of their organisation, societies do not exist in some immaterial abstract ether. Whatever impacts and interferences have already been produced, natural conditions form the ineliminable material basis for any form of society.

If humankind is living through a planetary crisis already underway, which will push the world into a state that is increasingly hostile to life, an idyllic society of material abundance and absence of scarcity is outside the scope of possibilities. And if it is outside the scope of possibilities, it has to be outside the imaginary that aims to mobilise the kind of social transformation advocated by the ecosocialist tradition.

Notes

1 Foster, "Review of environmental politics" (2000a).
2 Burkett, "Two stages of ecosocialism?" (2006).
3 Foster, *Marx's ecology* (2000b).
4 Fernandes, "Ecosocialism from the margins", 2020.
5 Foster, "Review of environmental politics" (2000a), 463.
6 I recommend Sabrina Fernandes' excellent presentation on this issue, entitled "Fundamentos do ECOSSOCIALISMO, número 049" ["Foundations of ECOSOCIALISM, number 049"] (2019). See www.youtube.com/watch?v=WcpZG3HkEtQ&list=PLPZ4y7b7MwOv0HV55UCMbXYFh6jjjneDP&index=2
7 O'Connor, "Contradictions of capitalism" (1991); O'Connor, "Capitalism, nature, socialism" (1988).
8 Burkett, "Two stages of ecosocialism?" (2006).
9 Ibid., 23–24.
10 Marx, *Capital*, vol. 1 (1982), 649–650.
11 Burkett, "Two stages of ecosocialism?" (2006), 35.
12 Marx, *Capital*, vol. 1 (1982).
13 England *apud* Burkett, "Two stages of ecosocialism?" (2006), 32.
14 White, Gareau, & Rudy, "Ecosocialisms, past, present and future" (2017).
15 Meadows, Randers, & Meadows, *Limits to growth* (2005).
16 White, Gareau, & Rudy, "Ecosocialisms, past, present and future" (2017), 28–29.
17 Ibid., 34–35.

References

Burkett, P. (2006). Two stages of ecosocialism? *International Journal of Political Economy, 35*(3), 23–45.

Fernandes, S. (2020). Ecosocialism from the margins. *NACLA Report on the Americas, 52*(2), 137–143.

Foster, J. B. (2000a). Review of environmental politics: analyses and alternatives. *Capital & Class, 72*, 461–477.

Foster, J. B. (2000b). *Marx's ecology*. New York: Monthly Review Press.

Marx, K. (1982). *Capital* (Vol. 1). London: Penguin Books.

Meadows, D., Randers, J., & Meadows, D. (2005). *Limits to growth: the 30-year update*. London: Earthscan.

O'Connor, J. (1988). Capitalism, nature, socialism a theoretical introduction. *Capitalism, Nature, Socialism, 1*(1), 11–38.

O'Connor, J. (1991). On the two contradictions of capitalism. *Capitalism, Nature, Socialism, 2*(3), 107–109.

White, D. F., Gareau, B. J., & Rudy, A. P. (2017). Ecosocialisms, past, present and future: from the metabolic rift to a reconstructive, dynamic and hybrid ecosocialism. *Capitalism, Nature, Socialism, 28*(2), 22–40.

9 Capitalocene and world-ecology

A theoretical alternative to the Anthropocene and metabolism?

As seen in Chapter 2, issues relating to the metabolic interaction between societies and nature occupy a prominent place in ecosocialist thinking. The theoretical tradition that has been formed and consolidated on the basis of reflection on these issues has also adhered to an emerging consensus in the natural sciences around the concept of the Anthropocene.

The first step to understanding this concept is to recognise that any form of society always presupposes a natural world, i.e., its existence and reproduction depend on a set of natural conditions. It is in the field of possibilities determined by this natural world that societies find the space and limits (some relative, some absolute) for their development.

More specifically, the development of civilisation as it is today was made possible by a geological epoch of relative climatic stability, the Holocene. Everything that characterises this time interval of just under 12,000 years (temperature levels and zones, winds, water and rainfall regimes, ocean levels, sea currents, the cryosphere, etc.) paved the material (biogeochemical) conditions for the historical processes that brought humankind to where it stands. However, especially since the first Industrial Revolution, disturbances in this relative stability have been triggered, affecting the various ecological parameters that give material support to societies. We have triggered environmental, ecological, and climatic changes. Changes that occur at rates that make it increasingly difficult to re-establish those relatively stable patterns. Changes that, as a result, produce increasingly frequent and intense stressors for the reproduction of life (in general) and societies (in particular) as they have evolved *so far*. This is what may be called the *ecological crisis*.

Even in the field of natural sciences, where social critique is hardly ever part of the agenda, there is already a clear understanding that human impact on the planet has started to show new characteristics in very recent times. As Angus reports, at a conference of the *International Geosphere-Biosphere Program* (IGBP) just 20 years ago, atmospheric chemist Paul Crutzen, irritated by the constant references to the Holocene, voiced, almost on impulse, that the word

DOI: 10.4324/9781003512516-11

was no longer appropriate and should be replaced by the term *Anthropocene*.[1] The term would indicate that humanity has become a relevant driver of global ecosystem transformation.[2] Crutzen was already a prestigious Nobel laureate in chemistry at the time, so it is not surprising that his outburst had consequences, triggering research efforts and an avalanche of publications around this new term.

Moreover, much more than Crutzen's prestige, the obvious adherence that the term has to reality is largely responsible for its impact on scientific activity. Subsequently, many periodisations have been suggested to mark the beginning of this new geological epoch, and, to date, the *International Commission on Stratigraphy* (ICS) and the *International Union of Geological Sciences* have still not approved the exact change in the geological time scale, such is the variety of moments in which the human mark could be detected stratigraphically. In 2019, the Anthropocene Working Group (AWG), commissioned by the ICS, issued an opinion with the formal proposal to place the start of the new geological epoch in the middle of the 20th century.[3]

This proposal is consistent with what has become known as the Great Acceleration. In 2004, a team of IGBP researchers published a collection of graphs that illustrated significant and simultaneous inflections in the trajectories of a series of social and natural variables from the 1950s onwards.[4] Population, gross domestic product (GDP), urban population, use of fertilisers, water consumption, transportation, carbon dioxide, methane, temperature, ocean acidification, etc., all shot upward from 1950 onwards. Therefore, Veiga says

> no one has the right to ignore the fact that, at least since the middle of the 20th century, humans have exerted excessive pressure on some of the most important biogeochemical cycles, such as carbon and nitrogen. At the same time, there has been an unprecedented general escalation of other artificial (anthropogenic) impacts on the biosphere. It may be enough to remember that, of all the carbon dioxide attributable to human activities stored in the atmosphere today, three-quarters have been emitted in the short period of the last seventy years.[5]

Jason Moore carefully recognises that this whole debate is circumscribed to the geological debate and stratigraphy. However, he rightly challenges the validity of this circumscription for the debate on the ecological impacts of capitalism. Indeed, the discussions about the Anthropocene involve finding clear signatures of specifically human impact in the geological strata. That is, the search is not for signs of our impact in a broad sense. The search is for signs of our impact that are recorded geologically. And when this is the focus of the search, Moore argues, a whole long ecological history of the emergence of capitalist society is simply brushed aside because its geological marks are not so indisputable.[6] To him, not even the first Industrial Revolution should be taken as the great turning point, but strictly speaking, the geological community is more likely to place this demarcation even closer to the present day, in the middle of the 20th century.

It is important to highlight that the focus of Moore's critique is not that, from a strictly geological point of view, the human imprint might need to be sought much further back than in the 19th or 20th century. Nor is he simply saying that the term Anthropocene places all of humanity, as a whole and in an undifferentiated way, as responsible for a set of impacts. This is indeed problematic, but the author is pointing out more than that. He maintains that, from the standpoint of the ecological critique of capitalism, the geological debate is hopelessly incomplete and insufficient. Everything that is necessarily left out of this search for specific marks in geological strata prevents the realisation, according to him, that the main driver of change on the planet has been, at least since the 16th century, capital.

Therefore, when proposing the term Capitalocene, he is not proposing a geological term to replace the Anthropocene. He is proposing a framing of the issue that escapes the boundaries of stratigraphy and enables the reformulation of the ecological critique of capitalism in such a way as to succeed in contemplating the whole range of planetary transformations driven throughout the historical emergence of this society, not just the transformations that are geologically marked.

In other words, the predominance of the term Anthropocene does not only concern a debate about demarcation on the geological time scale. It is a predominance that *reveals and organises* a whole way of thinking about the interaction between societies and nature. And the first aspect of this way of thinking (both in science and outside it) to which Moore draws attention is what he calls Green Arithmetic. Other than this, there are two other important aspects of his argument to be discussed: his concept of world-ecology and his discussion of the history of capitalism.

What he means by Green Arithmetic, in summary, is a kind of perspective that takes natural reality and social reality as separate things. On the one hand, pure nature, and on the other, society, endowed with complete autonomy. When the sciences of society consider environmental issues, they perform this arithmetic by adding nature *a posteriori*. When the natural sciences deal with social issues, they perform a similar operation, grafting society. The verb "graft" is not employed here by chance. This passage from Moore's critique, which he calls Green Arithmetic, bears some similarities to some of the critiques that rift scholars have levelled at first-stage ecosocialist thinking.

It may seem strange to mention the possibility of the natural sciences contemplating social issues. This strangeness is quickly dispelled by remembering that, in a scenario of multiple environmental collapses, natural scientists have increasingly been forced to contemplate the causes, drivers, and social implications of the phenomena they seek to understand. Being rigorous with Moore's reflection, this entanglement between natural and social is typical of our reality, not of a specific moment or phenomenon. On the other hand, it is also true that this entanglement is revealed (exhibited) more clearly in certain phenomena, which demonstrates how problematic the typical perspective of Green Arithmetic is.

An interesting way of highlighting this problematic nature is to take a step forward in the author's discussion of the exceptional nature of humanity. According to Moore, a way of thinking that takes humanity as an absolutely unique species is organised around the notion of the Anthropocene. Logically, Moore does not disagree, and nor could he, with the existence of human exceptionality. However, he is careful enough to ask the following question: exceptional, singular, unique, in what sense? He defends that the Green Arithmetic perspective conditions the answer. And in this case, it leads to the wrong answer.

He states:

> Seeing human relations as not only distinct from nature, but as effectively independent of the web of life, has shaped social thought for two centuries. [...] In this, human exceptionalism expresses the peculiar idea that humanity "alone is not a spatial and temporal web of interspecies dependencies" [...] In this view, humans are recognized as one species within the web of life (One System). But the recognition proceeds by abstracting – rather than synthesizing – the biological from human sociality. [...] Practically speaking, Society is independent from Nature (Two Systems). For the earth-system scientists behind the Anthropocene, Social Factors [...] are added; for scholars in the humanities and social sciences, Nature is added.[7]

It is not hard to imagine how such an understanding may be capable of supporting ideas such as post-scarcity, discussed in the previous chapter.

By combining these two discussions, Moore's critique becomes more powerful. It demonstrates that the mistake of assuming scarcity is a purely social phenomenon is firmly rooted in a weakness of origin, i.e., in a problematic ontology. To put it more simply, what Green Arithmetic does is very broadly organise a certain worldview, a certain perspective on both natural and social reality. And this worldview informs conceptions that say it is possible to overcome any scarcity just by reorganising society. It is truly disturbing that serious and critical intellectuals dedicated to thinking about the main ecological issues of our time may fall into such a trivial error. They not only overlook the interdependencies that exist between nature and society. They also ignore the fact that no society, whatever it may be, can freely build nature as it chooses.

In other words, resorting to Burkett, although society (whatever it may be) develops in coevolution with nature, it does not do so under freely chosen conditions.[8] This also echoes a well-known passage by Marx in his *The Eighteenth Brumaire of Louis Bonaparte* in which he states that human beings "make their own history, but they do not make it just as they please; they do not make it under circumstances chosen by themselves, but under circumstances directly encountered, given and transmitted from the past."[9]

In short, already at this point, Moore's critique maintains, which should come as no surprise, that if the theoretical foundations are weak, the theoretical and practical implications will be fragile. But it is worth going even further than this conclusion. Recalling the last chapter, Marx's critical endeavour is not satisfied with pointing out the limits and problems of a theory. Any theoretical demolition is *always* followed by the question: how can a theory that is so clearly problematic not be discarded? Why does it circulate, being reproduced even by scientists and intellectuals with a sincere ambition to understand reality?

In this specific case, Marx himself points the answer, in a passage from the *Grundrisse*, much quoted by Foster:

> What requires explanation is not the unity of living and active human beings with the natural, inorganic conditions of their exchange of matter with nature, and therefore their appropriation of nature; nor of course is this the result of an historical process. What we must explain is the separation between these inorganic conditions of human existence and this active being, a separation which is posited in its complete form only in the relationship between wage labour and capital.[10]

Recognising that this statement is correct, one must also recognise that Green Arithmetic does indeed reflect (albeit in an uncritical and ahistorical way) a characteristic that is specific to capitalist society. On the one hand, this sheds light on the real basis that gives social support to this mystified standpoint. On the other hand, it explains why this standpoint has such a limited critical horizon: *it accepts what is specific to capitalism as something specific to human life in society*. Therefore, if society and nature are irremediably separate realities, there are only two paths for any social organisation: either dominate nature or succumb to it.

Taking the opportunity to return to Moore's text, recognising that a mystified idea has a real basis does not imply concluding that it adequately captures that real basis. That is, while admitting that Green Arithmetic inadvertently reflects a real separation between society and nature, one may point out that it does so incorrectly. This is because such a separation is based on an unavoidable process of coevolution between society and nature. Moore seeks to capture this coevolutionary dimension by countering Green Arithmetic with his concept of world-ecology.

Immediately, the question of limits arises again. According to Moore, the worldview informed by Green Arithmetic provides an open flank for neo-Malthusian perspectives. On this point, as well as on the one about scarcity, Moore does considerably better than White and his co-authors, discussed in Chapter 8. He is quite clear that the problem with neo-Malthusian ideas is *not* that they emphasise population. The problem, beyond their political and social implications, is that they start from a conception of population dynamics that is completely independent of social reality (in general) and the development of capitalist society (in particular). That said, it is easy to see how Green Arithmetic sets the stage to notions like this.

The concept of world-ecology challenges not only the impervious separation of nature and society. It also challenges any notion of unidirectionality. In other words, the notion of interdependence that Moore proposes is such that the dynamics of capital interfere with these relationships and are also, to some extent, shaped by them. The importance of this cannot be underestimated. It is relatively simple to dismiss theories that take nature as something absolute and society as a mere human grouping oppressed by the inclement forces of natural limits. However, at least according to Moore, traditions that seem to escape Green Arithmetic end up falling into the same error, but in the opposite direction.

His main target, to the surprise of many, is the rift scholarship (in general) and John Bellamy Foster (in particular). According to Moore, the notion of metabolic interdependence in this tradition overlooks nature more than it should. This may sound a little strange, but considering the other side of the coin, the author's criticism starts to seem more plausible. What he is saying is that the tradition of metabolism over-autonomises society. And this has a decisive implication, which Moore explicitly points out: if society's autonomy from nature is overrated, the conception of emancipatory strategies acquires an illusory range of alternatives. In other words, it seems to benefit from a freedom of action that does not actually exist.

In the following passage from the text, the author explains this understanding in a very effective way:

> Such views evidently rest upon Human/Nature dualism and its cognates. This dualism obscures our vistas of power, production and profit in the web of life. It prevents us from seeing the accumulation of capital as a powerful web of interspecies dependencies; it prevents us from seeing how those interdependencies are not only shaped by capital, but also shape it; and it prevents us from seeing how the terms of that producer/product relation change over time. For instance, it is clear that capitalogenic climate change is undermining crucial relations of capitalism's Cheap Food regime in the twenty-first century – Cheap Nature increasingly confronts forms of nature that cannot be controlled by capitalist technology or rationality.[11]

The end of this excerpt mentions something important: the notion of Cheap Nature. Before diving into it, a brief clarification. Moore uses the term "capitalogenic" to describe climate change. In the same way that the established term "anthropogenic" refers to humanity as the main driver of these changes, the term "capitalogenic" is intended to replace it, indicating capital as the real main driver.

Resuming the argument, Moore takes advantage of the categories of *organic composition* and *rate of profit* to shed light on the important interdependence between the dynamics of capital and nature. Marx presents the organic composition of capital in Chapter XXIII of Book I.[12] This category informs the relationship between constant capital (which takes the form of means of production) and variable capital (which takes the form of labour-power). At the time, Marx was more interested in showing how accumulation tends to slow down if capital cannot cheapen labour-power on the one hand and do without it on the other. This is achieved by increasing/raising/enhancing productive forces, which tends to increase/raise/enhance the organic composition of capital.

Moore highlights that a similar process can be observed through the lens of the material preconditions of production. The course of accumulation does not only exert pressure on the availability of labour-power. On the one hand, as the scale and scope of production grow, so does the demand for means of production in general. In part, these means of production are raw materials and auxiliary materials, which tend to become more expensive. In part, they are machines, instruments, buildings, etc., the production of which also depends on mineral, energy, water, and other resources. Furthermore, the incorporation of a growing mass of workers also has repercussions on the increasing demand for food, natural fibres, etc. Moore's insight is that these growing demands on diverse resources *also put the brakes on accumulation*. As such, they must also be counterbalanced by the cheapening of these resources. The conclusion is that the logic of capital only becomes materially viable over time if it manages to create what he calls Cheap Nature.

This may seem trivial at first glance, but there is a crucial implication here, connected with a theme discussed in the previous chapter. Moore's theoretical argument shows that when capital faces material obstacles, i.e., the exhaustion of the material resources necessary for its operation, it actually *speeds up even more*. No brake is built into these processes of depletion. The race to cheapen what is already used becomes even more frantic, on the one hand, and the search for alternatives intensifies on the other. Material obstacles do not hold back the destructive tentacles of capitalist production. On the contrary, they penetrate even deeper and reach even further.

The point here is precisely the same as the one made by the critique that second-stage ecosocialists direct at James O'Connor's two contradictions. O'Connor seems to adhere to the idea that capital would eventually face a natural limit beyond which it would be forced to operate sustainably.[13] Taking a different route, Moore also maintains that this understanding is mistaken.

Based on this argument presented by Moore, one may ask the following questions: what kind of capitalism tends to be propelled in a world beset by a hostile nature? In a collapsing world, how long could a dystopian capitalism remain materially and dynamically viable?

Moore attempts to demonstrate how capitalism has historically created this Cheap Nature. The main argument he defends, it seems, is that to situate the opening of an era of ecological crises only from the first Industrial Revolution, as those who adhere to the notion of the Anthropocene usually do, is wrong. According to the author, capital has triggered several moments of accelerated ecological impacts since the 16th century.

Despite its relevance, this question will not be approached in detail here. At this point, it is more interesting to use the discussion about Cheap Nature as a hook for the next chapter. It presents two of the best known critiques levelled at Marx. The first one is common in any ecological debate involving Marxists: the claim that Marx is a thinker who is unconcerned about ecological issues because his theory of value only takes human labour into account, disregarding the participation of nature in the formation of wealth. The second relates to a productivist standpoint that Marx would occasionally show when addressing the relationship between social relations and productive forces.

Notes

1 Angus, *Facing the Anthropocene* (2016).
2 Veiga, *O Antropoceno e a ciência do Sistema Terra ["The Anthropocene and the Earth System science"]* (2019).
3 AWG, *Newsletter of the Anthropocene Working Group* (2019).
4 Steffen *et al.*, "Trajectories of the Earth System" (2018).
5 Veiga, *O Antropoceno e a ciência do Sistema Terra ["The Anthropocene and the Earth System science"]* (2019), 27.
6 Moore, "The Capitalocene, Part I" (2017).
7 Ibid., 596–597.
8 Burkett, "Two stages of ecosocialism?" (2006).
9 Marx & Engels, *Collected works*, vol. 11 (1980), 103.
10 Marx & Engels, *Collected works*, vol. 28 (1986), 413.
11 Moore, "The Capitalocene, Part I" (2017), 598.
12 Marx, *Capital*, vol. 1 (1982).
13 O'Connor, "Two contradictions of capitalism" (1991).

References

Angus, I. (2016). *Facing the Anthropocene: fossil capitalism and the crisis of the Earth System.* New York: Monthly Review Press.

Anthropocene Working Group. (2019). *Newsletter of the Anthropocene Working Group: Volume 9. Report of activities 2019.* Retrieved from http://quaternary.stratigraphy.org/wp-content/uploads/2020/09/Anthropocene-Working-Group-Newsletter-Vol-9-final.pdf

Burkett, P. (2006). Two stages of ecosocialism? *International Journal of Political Economy, 35*(3), 23–45.

Marx, K. (1982). *Capital* (Vol. 1). London: Penguin Books.

Marx, K., & Engels, F. (1980). *Collected works* (Vol. 11). New York: International Publishers.
Marx, K., & Engels, F. (1986). *Collected works* (Vol. 28). New York: International Publishers.
Moore, J. (2017). The Capitalocene, Part I: on the nature and origins of our ecological crisis. *The Journal of Peasant Studies, 44*(3), 594–630.
O'Connor, J. (1991). On the two contradictions of capitalism. *Capitalism, Nature, Socialism, 2*(3), 107–109.
Steffen, W., Rockström, J., Richardson, K., Lenton, T., Folke, C., Liverman, D., … Schellnhuber, H. (2018). Trajectories of the Earth System in the Anthropocene. *The Proceedings of the National Academy of Sciences, 115*(33), 8252–8259.
Veiga, J. E. (2019). *O Antropoceno e a ciência do Sistema Terra.* São Paulo: Editora 34.

10 A supposedly anti-ecological and productivist Marx

The discussion of cheap nature points to a way to respond to the critique of the supposed absence of nature in the Marxian theory of value. It is capital itself that needs a nature that is increasingly devoid of value, and that, *out of necessity*, actively produces an increasingly disposable world for itself. In other words, this alone suggests that this critique levelled at Marx must be levelled at capitalism itself. When it overlooks nature, Marx's theory of value captures an existing aspect of the social reality governed by capital. From this, this argument goes a few steps further.

In Chapter I of *Capital* (precisely when he presents the foundations of his theory of value), Marx shows that human life is constantly crossed by wants, by needs. Some of these needs can be directly satisfied by what is available in nature. However, the fulfilment of many of these needs requires the mediation of useful labour. Therefore, as a creator of use values, labour is a condition of human existence. In other words, it is a historically transversal feature of human life. Consequently, it is not a peculiarity of any specific social formation.

But what does it mean to say that meeting needs requires the mediation of labour? What exactly does labour mediate? Labour mediates between what nature offers as natural matter and the purpose set by those who work. Thus, use value, says Marx, is the nexus of two elements: natural matter and labour. According to the author, disregarding all the work involved in the formation of use value, one is left with a material substrate that exists in nature without any intervention by labour.[1]

But why is this important? While the only source of value is labour as abstract labour, the source of use value is labour as useful, concrete labour, *but also nature*. For this reason, Marx says, referring to William Petty, that as far as *material* wealth is concerned, labour is the father and the earth is the mother.

Burkett is an author who responds competently (and in several texts) to this specific critique. In the text discussed in this book, he emphasises that recognising nature as the material condition of value and surplus value *is not the same* as saying that nature makes up value. Like it or not, it *does not make up* value.

DOI: 10.4324/9781003512516-12

Furthermore, capital rushes to freely appropriate any natural or social element or force that has not yet been transformed into a commodity. That is what the tragedy of the commons captures in a mystified way.

A passage from Burkett helps to better understand what this free appropriation means:

> In Marx's analysis, the value of labor power (which regulates wages) only includes the value of the commodities entering into labor power's reproduction. Obviously, capitalists do not pay for the noncommodity elements of this reproduction, including not only domestic labor but also fresh air, sunlight, and rest time. Rather, they freely appropriate these noncommodity "reproduction costs" in the form of the surplus wage labor that they enable workers to expend.[2]

In this passage, Burkett echoes one of the most important theoretical elements of a part of Marxist feminism known as the theory of social reproduction. It is easy to see that the reasoning is equally valid for any aspect of nature that is not yet commodified. By highlighting the voracious appropriation of all the elements of life (natural and social) that can drive capitalist accumulation, Burkett also sheds light on a significant connection between a possible Marxist ecological thought and Marxist feminism.

The second critique of Marx points to a supposedly productivist character in his thinking, which would indicate an adherence, at least at times, to a notion of humanity conquering nature. One of the best-known ecosocialists to convey this type of critique in several of his texts is Michael Löwy.

First of all, Löwy's critique *does not* coincide with accusations of productivism coming from non-Marxist traditions of ecological thought. In fact, in his famous essay "What is ecosocialism," Löwy states:

> Ecologists accuse Marx and Engels of productivism. Is this justified? Yes and no. No, to the extent that no one has denounced the capitalist logic of production for production's sake – as well as the accumulation of capital, wealth, and commodities as goals in themselves – as vehemently as Marx did. The very idea of socialism [...] is that of production of use values, of goods necessary to the satisfaction of human needs. For Marx, the supreme goal of technical progress is not the infinite accumulation of goods [...] but the reduction of the working day and the accumulation of free time.[3]

However, once he has admitted that Marx is not the productivist thinker that ecologists in general think, Löwy goes on to try to show that, on the one hand, both Marx and Engels would often give in to the temptation to positively evaluate the supposed civilising nature of capitalist production and the growing capacity to instrumentalise nature; on the other hand, they seemed to ignore the possible existence of natural limits. In other words, although not saying it outright, it appears that, for Löwy, Marx was only critical of productivism when it was explicitly capitalist.

Several passages from various works are brought up by the author to reinforce his point. Some could certainly sway the reader to the side of Löwy's critique. Without disregarding the fact that these passages, and perhaps many others, can actually be interpreted the way the author does, another kind of exercise is possible, trying to offer what is contradictory.

Löwy quotes Marx from the Foreword to his 1859 work, *A Contribution to the Critique of Political Economy*:

> At a certain stage of development, the material productive forces of society come into conflict with the existing relations of production [...]. From forms of development of the productive forces these relations turn into their fetters. Then begins an era of social revolution. [...] No social order is ever destroyed before all the productive forces for which it is sufficient have been developed.[4]

Löwy claims that in this passage productive forces appear as neutral and that the task of the revolution would only be to abolish capitalist relations and thus suspend the obstacles to the development of productive forces. Reading the passage quoted by the author again, however, it is clear that there is nothing neutral about productive forces. The contradiction between productive forces and relations of production only exists because the productive forces are not neutral. And at a stage of their development, this contradiction becomes insurmountable. Productive forces and the relations of production generating them become irremediably incompatible. At its climax, this contradiction can only be resolved with the collapse of such relations. But in this passage, absolutely nothing qualifies this "epoch of social revolution" as a communist revolution. And even less is said about its task or desired results.

After that, Löwy quotes a slightly longer passage, this time from the *Grundrisse*:

> Thus, just as production founded on capital creates universal industriousness on one side – i.e. surplus labour, value-creating labour – so does it create on the other side a system of general exploitation of the natural and human qualities [...] Thus capital creates the bourgeois society, and the universal appropriation of nature as well as of the social bond itself by the members of society. Hence the great civilising influence of capital; its production of a stage of society in comparison to which all earlier ones appear as mere local developments of humanity and as nature-idolatry. For the first time, nature becomes purely an object for humankind, purely a matter of utility; ceases to be recognized as a power for itself; and the theoretical discovery of its

autonomous laws appears merely as a ruse so as to subjugate it under human needs, whether as an object of consumption or as a means of production.[5]

In this passage, Löwy sees an uncritical admiration for the civilising domination of capital over nature. Again, it seems fair to redirect this critique of Marx towards capitalism. Furthermore, nothing in this passage suggests that Marx makes a positive value judgement of the processes described.

The only sentence that might support Löwy's reading is the one in which Marx affirms the great civilising action of capital. However, Löwy might have fallen victim to a semantic slip similar to the one that sees the terms "development" or "evolution" as a passage from a worse state to a better one. Marx did not use any of these terms in this positivist register. On the contrary, he repeatedly displayed full awareness that processes of development or evolution could well be worthy of negative value judgement. He is also fully aware that such civilising action can be populated with tragedy or even be predominantly tragic.

Since Löwy has apparently based his reading of this text on a single word, it seems appropriate to offer a counterpoint that rests on a single word. In the last sentence quoted, Marx says: "the theoretical discovery of its autonomous laws appears merely as a ruse so as to subjugate it under human needs, whether as an object of consumption or as a means of production," Okay, note that the word "ruse" carries many meanings indicative of a negative value judgement. The word points to an idea of manoeuvring to deceive, of a cunning scheme, of a subterfuge. In the sentence quoted, the cunning scheme is precisely oriented towards submitting nature to human needs.

This exercise could go on for longer. However, so as not to make this chapter more tedious than necessary, there are at least two lessons to be learned from the two critiques analysed here. First, Marx's thinking *is not normative*. It is not directed at a world that he would like to exist, as he would like it to exist. The object of Marx's reflection is very clear. It is capitalist society. If the characteristics revealed by his research seem reprehensible, the question is whether Marx was wrong or *whether it is reality itself that is reprehensible*. One should not try to find in Marx elements of a world that should be. The second lesson is that no form of society arises from thin air. Every new social formation always rises from the rubble of the previous one. Recognising this is, at the same time, admitting that an eventual ecosocialist transition to communism will necessarily have to be constituted from the world left behind by capital. Such recognition does not imply assuming that productive forces are neutral, for example. They obviously are not. But it is from them that a world beyond capital will begin to be built.

Notes

1 Marx, *Capital*, vol. 1 (1982).
2 Burkett, "Two stages of ecosocialism?" (2006), 27.

3 Löwy, *What is ecosocialism?* (2005), 2.
4 Marx *apud* Löwy, *O que é o ecossocialismo? ["What is ecosocialism?"]* (2014), 25–26.
5 Marx *apud* Löwy, *O que é o ecossocialismo? ["What is ecosocialism?"]* (2014), 26–27.

References

Burkett, P. (2006). Two stages of ecosocialism? *International Journal of Political Economy, 35*(3), 23–45.
Löwy, M. (2005). What is ecosocialism? *Capitalism, Nature, Socialism, 16*(2), 15–24.
Löwy, M. (2014). *O que é o ecossocialismo?* São Paulo: Cortez.
Marx, K. (1982). *Capital* (Vol. 1). London: Penguin Books.

Part III

Tactical and strategic issues

11 A crisis to subordinate all others

An overview of climate collapse

Writing about climate change imposes some rather peculiar difficulties. The object of study changes very quickly. Examples of extreme events become outdated quickly. They are soon overtaken by worse, more numerous, or more frequent events. The data also changes all the time and soon becomes outdated. Mitigation trajectories and targets need to be revised frequently. Forecasts need to be revised frequently. Models are continually being modified.

Much of the information in this chapter comes from the first time this material was prepared, between 2009 and 2012. Updates and small observations have been added as needed. However, the changes just mentioned are all refinements of the understanding of the phenomenon that point in the same direction: the increasing destabilisation of the climate system. Having this in mind, the data and information presented are enough to build this argument.

The main environmental challenge humanity is facing today is the phenomenon known as climate change. First of all, it is important to get used to some basic concepts in this field of research, a task that economists do not always tackle.

Most people have heard of *greenhouse gases* and know, even vaguely, the effect they have. This chapter will present some additional details on this concept. On our planet, there is a delicate energy balance between the energy coming from the sun, as shortwave radiation, and the energy radiated by the planet itself into space, as longwave radiation. Greenhouse gases are "transparent" to shortwave radiation, so energy coming from the sun passes right through them. However, they interact with longwave radiation, reflecting back to the planet part of the energy that would otherwise be radiated into space. This fraction of longwave radiation is in the infrared portion of the spectrum, associated with heat, as known since the turn of the 18th to the 19th century.[1]

Greenhouse gases are present in atmospheric chemistry in tiny percentages. But it is enough to alter the planet's climate substantially. Without them, it is estimated that the average temperature would be around minus 18°C, that is, over 30°C colder than it is today.[2] Given these proportions, it is not surprising

DOI: 10.4324/9781003512516-14

that even slight variations in their quantities in the atmosphere can cause significant variations in the planet's average temperature.

Therefore, the *concentration of these gases* is the second important concept this chapter addresses. One of the commonly used measures is the CO_{2e} concentration, which converts the radiative forcing of all greenhouse gases into equivalents of the radiative forcing of carbon dioxide, CO_2. In simpler terms, it counts the concentration of all greenhouse gases in terms of their equivalence to CO_2. But equivalence in what sense? In their ability to trap longwave radiation, which warms the planet.

During most of the geological epoch we call the Holocene, the concentration of these gases in the atmosphere was around 280 parts per million. This metric means that there are 280 molecules of greenhouse gases in 1 million molecules of air. Since the Industrial Revolution, at the end of the 18th century and the beginning of the 19th century, this concentration has persistently increased. In 2010, according to the European Environment Agency, this figure had already risen to 431 ppm. According to estimates, this was the highest concentration level in the last 420,000 years. But in 2012, the concentration reached 439.9 ppm, which, according to the Intergovernmental Panel on Climate Change, the IPCC, would put us at the highest level for 800,000 years. But it did not stop there.

In 2020, the US National Oceanic and Atmospheric Administration reported that the concentration of atmospheric CO_2 (*only* CO_2) had reached 409.9 ppm, an unprecedented mark for the last 3 million years. At the same time, using different techniques for reconstructing paleoclimate data, an article published in the journal of the Geological Society of America estimated that in the troposphere, which is the first layer of the atmosphere, 10–12 km thick, the current concentration of CO_2 is unparalleled over the last 7–23 million years.

Of course, it must be considered that the variability of paleoclimate data is relatively high. Different teams obtain different estimates using different methodologies. Even so, what may seem insignificant from a purely numerical point of view is launching humankind into a world with an atmosphere that our species has never seen before. From a geological point of view, which considers the variations and changes over aeons, eras, and epochs, the transformations underway now have a fulminating speed.

What has caused this accelerated change? As most people know, anthropogenic greenhouse gas emissions, i.e., the emissions generated by our activities. There are ways of knowing for sure that the excess carbon in the atmosphere has our signature, by checking the ratio between the isotopes of carbon 12 (predominant as a residue from burning fossil fuels) and carbon 13 (predominant in natural emission processes).[3] But that is not exactly the point here, since the consensus around the anthropogenic nature of climate change is increasingly solid.

The point is the *fundamental distinction between emission and concentration*. Although this may seem like a trivial exercise, many people conflate these concepts. Anthropogenic emissions are the waste product of human activities, such as energy generation, industry, transportation, agriculture, and more. Since 2003, humanity has exceeded 40 billion tons of CO_{2e} emitted annually. And since 2017, our emissions have reached more than 50 billion tons of CO_{2e}. Not all of this staggering volume ends up in the atmosphere. According to the IPCC, between 1750 and 2011, oceans and vegetation absorbed an average of just under 5 billion tons of CO_2 each year. The surplus accumulates in the atmosphere. It is this surplus that will cause increases in the concentration of greenhouse gases. Thus, from the point of view of human interference, the concentration is the result of past (and present) emissions accumulated in the atmosphere.

This is why this relationship is important. The following illustration helps to understand it. In 2020, the Covid-19 pandemic took the world by storm. As a result of policies to contain the virus, industrial activity paused, especially in China. With urban populations trapped at home in several countries, the circulation of cars and other motor vehicles drastically reduced. Air traffic was also largely suspended. All of this contributed to an unprecedented drop in our emissions in 2020. Some of the most optimistic estimates mention a drop of up to 8%. Not paying attention to the difference between emissions and concentrations could lead to the mistaken conclusion that this drop in emissions halts, at least momentarily, the deepening of the problem. However, according to NASA data, between March and November 2020, the concentration of CO_2 (only CO_2) in the atmosphere increased from 413 ppm to 415 ppm.

It is not hard to see why. Supposing a level of emissions of 50 billion tons of CO_2 and applying the *most optimistic* estimate of a reduction in emissions in 2020, there would have been 46 billion tons of CO_2 emitted in that year. If oceans, forests, and vegetation in general absorb around 5 billion tons, this means that even with the unprecedented drop in global emissions, there would still be 41 billion tons of CO_2 launched into the atmosphere.

Therefore, as long as human emissions are above the level that oceans and vegetation can absorb, the concentration of greenhouse gases in the atmosphere will continue to rise. Furthermore, this absorption capacity is not fixed. Some authors even claim that it can undergo drastic and abrupt variations. Apart from this, it is evident that deforestation reduces this capacity. In addition, as the oceans absorb CO_2, they become more acidic, destroying phytoplankton populations, the very organisms that are largely responsible for the oceans' ability to absorb more CO_2 than they emit. This process of ocean desertification could even turn the oceans into net emitters of CO_2.

And what are the *expected impacts* of climate change? Until recently, there were two main lines of interpretation, which differed fundamentally in their understanding of what is called the *tipping point*, the point beyond which changes completely escape the possibility of conscious and targeted intervention. Until the early 2000s, the dominant standpoint was well represented by the IPCC reports. Considering the concentration of greenhouse gases and the average temperature of the planet to be in a stable relationship, the tipping point was understood in terms of temperature. Beyond a certain tipping point, the temperature would spiral out of control due to feedback mechanisms that would autonomously drive up the concentration. Still, even when crossing the tipping point, the relationship between temperature and concentration would remain relatively stable.

The second line of interpretation, known today as Earth System theory, emphasises the biogeochemical interactions that make up the planet's natural systems. It does not assume that the relationship between the planet's average temperature and the concentration of greenhouse gases is necessarily stable. This completely changes the character of the tipping point. Rejecting the idea of a stable relationship between temperature and concentration, it admitted that, from the tipping point onwards, the temperature could continue to rise dramatically even if the concentration level became stable (or even declined). This line of interpretation has been corroborated for many years by paleoclimate evidence and has now become predominant, having even been incorporated into the most recent IPCC reports.

The notion of a tipping point has been presented here in a simplified way: in the singular. Today, it is known that there are countless tipping points scattered throughout the planet's natural systems. Some of them have probably already been crossed. As these tipping points are crossed, more fuel is added to the fire of the rise in the average temperature of Earth.

This is why, to this day, so many international conventions have been organised to set targets to prevent this warming from exceeding a certain level. At the beginning of the 1990s, the UN stated that an average temperature rise of 1°C above pre-industrial levels should be avoided. In the early 2000s, following a study commissioned by the British Crown, the most common goal became not to exceed warming of 2°C.[4] And since the mid-2010s, the most common target has been to limit warming to 1.5°C above pre-industrial levels.[5]

Furthermore, the impacts to avoid are numerous and the set of predicted impacts increases over time. Drought and desertification in Africa, Australia, southern Europe, and the western United States with *significant loss of arable land*. Glacial loss and melting of the polar ice caps. Rising ocean levels. Mass extinction of plant and animal species.

It would be a mistake to believe these effects are expected in the distant future. On the contrary, what is already happening has exceeded some of the most pessimistic forecasts. In 2007, an article published by the prestigious journal *Science* showed that sea levels were rising 60% faster than the IPCC projections. The temperature, meanwhile, was rising 30% faster.[6] The process of drought and desertification is taking place all over the world, and the historic drought in Syria in the early 2010s is an example. There is the constant northward displacement of arable areas in the United States. The melting observed in the Arctic in 2007, 2016, and 2019 was expected to only happen in the middle of this century. What happened in 2012 and 2020 was only expected for the end of the century. A 2019 UN report estimates the extinction of 250 species a day. At the time of the final preparation of this book, we have observed the global temperature record for each month being broken for 12 consecutive months, something absolutely exceptional and unprecedented. Even more impressive, the same persistent pattern of record-breaking has been occurring for ocean temperatures, but for 15 consecutive months. Given this scenario, how are mitigation strategies being formulated?

The most common thing the lay public comes across is temperature targets. Everyone is familiar with the goal of not exceeding an average temperature rise of 2°C above pre-industrial levels. And many probably also know that, since the Paris agreements, a target of 1.5°C has been favoured.

Before commenting on this type of target and the strategies that go with it, it is worth commenting on a topic that was more prominent in the first decade of the 21st century than it is today: The debate around CO_{2e} concentration targets.

One of the lowest targets ever proposed is that of 350 ppm. It was suggested by James Lovelock, an English scientist who was one of the forerunners of the Earth System theory, briefly mentioned at the beginning of this chapter. This target was formulated taking into account the idea of a tipping point. Thus, we can infer that, at least for Lovelock, from 350 ppm onwards, the trajectories of greenhouse gas concentration and average temperature of the planet would become decoupled, accelerating the temperature rise, even if the concentration became stable.[7] However, it is worth pointing out that the last time the CO_{2e} concentration was below 350 ppm was in 1977. Meeting this target would, therefore, require negative global emissions, i.e., removing more greenhouse gases from the atmosphere than emissions add to it. This topic will be further explored in due time, as there has been a lot of debate on the idea of negative emissions recently.

In any case, it is not hard to see why the 350 ppm target has remained a marginal, albeit rigorous, proposal from a scientific point of view. The most common target oscillated between 450 and 550 ppm. It could be found both in the IPCC reports and in the famous study *The Economics of Climate Change*, commissioned by the British Crown.[8] When these targets began to circulate in 2006 and 2007, the CO_{2e} concentration was 420 ppm. In 2018, concentration reached 456 ppm. Considering a recent growth rate of 3–4 ppm per year, it becomes clearer why the idea of net negative emissions has suddenly taken centre stage. Given the relationship between concentration and emissions, any hope of curbing the escalation of this chemical change in the atmosphere requires significant emissions mitigation trajectories.

In 2007, Nicholas Stern and his co-authors estimated the trajectories needed to meet the 450 ppm and 550 ppm targets as follows. To limit concentration to 450 ppm, the peak of the emissions would have to be between 2017 and 2018.

This means that, from then on, global emissions would have to decrease every year. But how much? According to the study, 5% a year until 2050. As a result, emissions in 2050 would be 70% lower than in 2007.[9]

There are two things to be observed in this proposal. First, this trajectory has already been made unviable by its suggested peak. Emissions continue to grow, and everything suggests that the year 2020 was just an exception. Second, at the beginning of the 21th century, it was still assumed that reaching 2050 with global emissions of 13.4 billion tons of CO_{2e} was enough. Today, there is talk of zero or negative emissions by 2050, and many climatologists have already warned that there is an urgent need to achieve this result in the 2020s.

In addition to this, there is the mitigation trajectory calculated for the 550 ppm target. In this case, emissions should peak between 2017 and 2027. Reductions after the peak should be between 1 and 3% per year. As a result, emissions in 2050 would be 25% lower than in 2007. In other words, in 2050, global emissions would reach 33.4 billion tons of CO_{2e} every year.

It should be clear by now that these trajectories are outdated. Given what is known today, they are patently insufficient. Even so, supposing that this would be a sufficient path to follow, this mitigation effort would face difficulties. Some emblematic success stories are enough to understand the challenge.

First, between 1977 and 2003, France massively reoriented its energy matrix towards nuclear energy, which has virtually zero emissions. Its electricity generation capacity from this source grew 40-fold, and its share of the French energy matrix exceeded 80%. Even so, in this period, France has managed to reduce its emissions by just 0.6% a year, which is insufficient even for the less demanding scenario mentioned above.

Then there is Brazil. Between 1975 and 2002, the share of biofuels in the transport sector increased from 1% to 25%. Did the sector's emissions reduce in the period? No. They grew by 2.8% a year.

Finally, there is the United Kingdom, which restructured its energy matrix on a similar scale to France, but towards natural gas. This led to a reduction in emissions of 1% per year between 1990 and 2000. Therefore, even a successful story like this would only be able to achieve the less demanding target of 550 ppm if *generalised to the whole world* and if emissions had started to decrease in 2017. Furthermore, the Hadley Centre, a research institute connected to the meteorological service of the United Kingdom, estimates that, with the atmospheric concentration of CO_{2e} at 550 ppm, there is a 99% probability of exceeding 2°C warming.

If this exercise, with outdated targets and trajectories, already reveals such a dramatic scenario, what is to be said about the more ambitious targets published since 2018? Since 2018, the IPCC's recommendation has been straightforward: arriving in 2030 with global emissions 50% lower than in 2010, i.e., 2030 global emissions cannot exceed 23.32 billion tons of CO_{2e}, and reaching at least zero net emissions by 2050.

Suppose the level of emissions in 2019 was the same as in 2018 (the last year for which data was available at the time of writing), i.e., 50 billion tons of CO_{2e}. Considering the biggest reduction estimated for 2020 due to the pandemic, 8%, that sets a starting point: 46 billion tons of CO_{2e}. To get to 2030 emitting only 23.32 billion tons, reductions of 7% every year until 2030 would be necessary. Almost the level of reduction that occurred as a result of the health and economic cataclysm of 2020 repeated every year for ten years.

However, the 2030 target is only intermediate. Achieving it is merely a condition to be in a position to reach the target that matters most, that of zero or negative emissions by 2050. Having achieved reductions of 7% per year starting from 46 billion tons, we would arrive in 2030 emitting 22.26 billion tons of CO_{2e}. By repeating this mitigation success for another 20 consecutive years, without ever failing, we would arrive in 2050 emitting 5.21 billion tons of CO_{2e}. Therefore, to achieve net zero emissions, it would be necessary to build up the capacity to extract those 5.21 billion tons directly from the atmosphere *every year*. To understand the magnitude of the technological feat required, it is estimated that all the oceans, all the forests, and all the vegetation on the planet absorb *less* than 5 billion tons of CO_2 per year.

The level of the challenge takes on a colossal magnitude when compared to what was thought and discussed less than 15 years ago. The scenario becomes even more acute considering that the IPCC estimates that the much-vaunted Paris Agreement would only take us halfway there even if there had been compliance with all the commitments signed.[10]

Given this scenario, what are the main strategies to mitigate CO_2 emissions proposed by traditional economic thinking? There are basically three: carbon pricing, energy policy, and removing barriers to behavioural changes (or creating incentives for virtuous behaviour).

These large groups of strategies aim at four determinants of anthropogenic emissions related to what has become known as the Kaya identity. In this identity, the level of emissions is given by the product of four factors: population, gross domestic product (GDP) *per capita*, energy intensity of GDP, and emissions intensity of energy consumption.

$$CO_2 \equiv pop \times \frac{GDP}{pop} \times \frac{E}{GDP} \times \frac{CO_2}{E}$$

The terms population, GDP, and energy intensity annul each other, leaving CO_2 equal to CO_2. This is why it is an identity.

A quick explanation: The energy intensity of GDP denotes how much energy needs to be consumed to obtain one unit of GDP. Energy intensity of GDP is often used as an approximate measure of energy efficiency. The lower the intensity, the higher the efficiency, and the lower the emissions. Emissions intensity of energy consumption denotes how much emissions each given volume of energy consumed generates. It is assumed that an energy matrix highly dependent on fossil fuels will have higher emissions intensity. Likewise, an energy matrix with a higher predominance of renewable energies will have lower emissions intensity.

Finally, given the intensities and the size of the economy, the larger the population, the greater the emissions. Likewise, given the intensities and the population, the larger the economy, the greater the emissions.

Considering everything covered in the introduction to this book, it is no coincidence that international climate policy has favoured the determinants of intensity instead of population and GDP. In both cases, energy policy is the strongest influence. If the policy is aimed at intensities, what are the determinants of these intensities?

Traditional literature points to three main factors: the proportion of energy-intensive and emissions-intensive activities in the national productive structure, the composition of the energy matrix, and the general level of energy efficiency.

Regarding the productive structure, the proportion that usually receives the most attention is that between agricultural, industrial, and service activities. It is usually claimed that there is a natural tendency for countries to follow a development path that starts cleaner, with agriculture predominating, goes through a period of increasing environmental impact, with industry predominating, and then moves on to a new low-impact phase, with services predominating. There are obvious problems with this type of formulation.

First, the assumption that agriculture has a low environmental impact. The degradation of the soil, the contamination of groundwater and local populations, the indiscriminate use of agrochemicals, and the emission of nitrous oxide through the use of fertilisers attest to the fact that large-scale agricultural activity, as it is today, has a tremendous impact. Nitrous oxide is actually a powerful greenhouse gas.

Second, the assumption that the service sector is a low-impact or even immaterial activity. Of course, this sector does not consume as much energy as heavy industry, for example. But what about electronic waste, its batteries, its heavy metals? What about air travel? What about the financial sector and its blackmailing practices that impose restrictions on any minimally serious environmental policy?

Third, the assumption that the higher predominance of the service sector leads to the shrinking of other sectors. This idea is completely unfounded. In a process in which the service sector becomes increasingly predominant, agriculture and industry may well continue to grow. Indeed, even if some countries have gone through a process of deindustrialisation, from a global point of view, all sectors are on an upward trend. Regarding climate change, what matters is not the ratio between activities but the scale on which they operate.

This was the first of the three main factors that determine energy and emissions intensities according to traditional literature: the productive structure. Next, there is the carbon content of the energy matrix.

The more fossil fuels predominate in the energy matrix, the greater the emissions intensity of that matrix tends to be. For this reason, a reduction in the carbon content of the energy matrix has been sought by encouraging renewable sources, which have an important role to play in global climate policy. However, a significant substitution is distant (to the point of not being foreseeable) not only in economic terms but also in technological terms. Data from the International Energy Agency (IEA) shows that between 1990 and 2018, the share of fossil fuels in the world energy matrix grew from 81.19% to 81.21%. How can this result be explained, given that these 28 years of international energy policy have been able to expand primary energy supply from renewable sources by 67%? It is enough to know that the primary energy supply from natural gas grew by 80% in the same period, that the supply from coal grew by 78%, and that the supply from oil grew by 38%. In absolute terms, the biggest increases in primary energy supply came from the following sources, from largest to smallest: coal, natural gas, oil, and renewables.

Finally, the third factor is raising the general level of energy efficiency. This is an achievable goal. More than that, it is a constantly achieved goal. More importantly, from the point of view of traditional economic thinking, it is a goal that is directly compatible with increasing levels of production and economic activity in general. However, as previously discussed in this book, all this success has been accompanied by apparently paradoxical results since significant increases in efficiency have not yet proved capable of curbing the expansion of energy consumption and greenhouse gas emissions. Even so, it is undeniable that stimulating energy efficiency has been the preferred instrument of energy policy focused on climate objectives.

Notes

1 Mann, *Climate change* (2020).
2 Ibid.
3 Ibid.
4 Stern, *Economics of climate change* (2007).
5 IPCC, *Global warming of 1.5°C* (2018).
6 Rahmstorf *et al.*, "Recent climate observations" (2007).
7 Lovelock, *The vanishing face of Gaia* (2009).
8 IPCC, *Climate change 2007* (2007); Stern, *Economics of climate change* (2007).
9 Stern, *Economics of climate change* (2007).
10 IPCC, *Global warming of 1.5°C* (2018).

References

Lovelock, J. (2009). *The vanishing face of Gaia: a final warning.* London: Penguin Books.

Mann, M. (2020). *Climate change: the science and global impact.* New York: edX/SDG Academy.

IPCC. (2007). *Climate change 2007: synthesis report.* Geneva: IPCC.

IPCC. (2018). *Global warming of 1.5 C: an IPCC special report on the impacts of global warming of 1.5 °C above pre-industrial levels.* Geneva: IPCC.

Rahmstorf, S., Cazenave, A., Church, J., Hansen, J., Keeling, R., Parker, D., & Someville, R. (2007). Recent climate observations compared to projections. *Science, 316*(5825), 709.

Stern, N. (2007). *The economics of climate change: the Stern review.* Cambridge: Cambridge University Press.

12 Geological voluntarism in ecosocialist thought

The discussion developed in this chapter establishes a critical dialogue with three aspects found in ecosocialist thinking. First, the overemphasis on the emancipatory and ecological content of the struggles of the working class. Chapters 5 and 8, on environmental ethics and the division of ecosocialist thought into phases, respectively, approached these issues. The final part of this chapter deepens the critique that can be directed at this over-dimensioning.

The second aspect concerns the expectation of incremental advances, i.e., small achievements that would accumulate in the course of ecological struggles. The impossibility of incremental gains has already been dealt with in enough detail in Chapters 4 and 11 on the myth of dematerialisation and climate change, respectively. This chapter explores a related issue concerning the debilitating character of unambitious struggles, which only aim at small advances and, therefore, act in a retail, pulverised way. This part builds on insights from Chapter 11 of *Capital*, in which Marx discusses cooperation.

The third aspect concerns struggles to *gain time*. The content from the previous chapter is enough to understand that the scale and speed of ongoing climate change, on the one hand, and the scale and speed of the necessary social transformations, on the other, make *any* claim that is content with merely gaining time *completely* unfeasible. Therefore, to avoid repetition, this chapter will not dive into the matter.

These preliminary notes already shed light on the necessity to address the demobilising potential of a reflection that leads to many blocked paths (precisely the paths that working-class struggles traditionally follow). This issue will be approached in the three following chapters.

Therefore, to avoid repeating discussions that have already taken place here (or are yet to take place), the structure of this chapter is leaner, exploring these issues in a slightly different way. It focuses on four points. First, a discussion of possibility and effectiveness. Second, a deeper look at what I call *geological*

DOI: 10.4324/9781003512516-15

voluntarism, in the light of the last four chapters. Third, the possibilities of working-class struggles. Finally, a discussion on cooperation and organisation.

For this first point, there is a shortcut based on some very fruitful insights from Aristotle.[1] The first informs us that

> even if one has a rational wish, or an appetite, to do two [...] contrary things at the same time, one will not do them; for it is not on this terms that one has the potency for them, nor is it a potency of doing both at the same time.[2]

Therefore, if the *means* do not conform to the *purpose*, the purpose is just an empty statement of intent. To put it in more specific and concrete terms, if the practices aim at preserving the conditions for the reproduction of capital, the aim of overcoming capitalism may only be realised through some unintentional historical stumble.

The second tells us that "then, which is capable of being may either be or not be".[3] This insight reinforces at least two specific and decisive moments in this book. When discussing the structural role of technology, it was defended that efficiency gains bring the *possibility* of saving resources, but this possibility is *not realised* in capitalism. When discussing market mechanisms, it was defended that some economic instruments or certain types of popular pressure have the *potential* to contain the environmental impact of activities but that this potential is only realised in limited and restricted spheres. This same intuition about the relationship between possibility and effectiveness supports the reflection on the working class' emancipatory (and even ecological) potential. The lesson learned is that recognising this emancipatory potential should not lead to mistakenly concluding that *any* struggle by the working class always necessarily points in the direction of emancipation.

The third states: "that which does not have the potency to be cannot exist anywhere".[4] The chapter on environmental ethics thoroughly discussed this point. For any purpose to be realisable, it needs to find in reality the conditions for its own realisation. Otherwise, as Lukács says, the purpose "is reduced to a fact of consciousness that becomes powerless in face of nature."[5] This is true for the simplest goal, and it is especially true for the most complex goals, such as the revolutionary transformation of social reality or the profound adaptation to climate change. It is not enough to conceive a desirable world for the future. It is necessary to find (and create) the objective conditions for achieving it in the present.

This connects to the second point to be addressed here, something I call geological voluntarism. The term describes an expectation, sometimes more explicitly formulated, sometimes more implied, that once humanity overcomes the destructive logic of capital, it would be within reach to establish the necessary metabolism between society and nature, capable of materially supporting an emancipated, sustainable, and even abundant society.

Usually, the very idea of abundance receives more attention. It is rightly said that the specifically capitalist abundance cannot be reproduced. Therefore, another kind of abundance must be sought: the abundance of care, of use, of duration, of what is necessary and not superfluous, of free time, leisure, etc. In short, an abundance of quality, not quantity.

However, it is of utmost importance not to lose sight of the fact that, just like the abundance of quantity, the abundance of quality *also rests on material prerequisites*. Not even this qualitative abundance can be sustained under *any* circumstances. *The planet inherited from the ruins of capitalism must be capable of sustaining this desired society.*

This is a point of contact with the rift scholarship and the critique levelled at it in Chapter 8. White, Gareau, and Rudy's article says that the rift scholarship ends up adopting a perspective that flirts excessively with the idea of equilibrium.[6] This appears in different intensities in the various authors of this tradition, but what is clear is that the emphasis on the metabolic rift is accompanied by an expectation that overcoming capitalism will bring the possibility of restoring the broken metabolism.

In its less sophisticated versions, this idea leads to the conclusion that if humanity has been able to completely destabilise the planet's natural cycles, it can also restore them. This is the epitome of what I call geological voluntarism because it assumes that it is within our power to consciously manage all the natural systems that make up the socio-ecological metabolism. By now, it should be clear that this assumption is wrong. In the same way that triggering an avalanche does not mean being able to reverse it (or even stop it), triggering some geological processes does not necessarily mean being able to set them in the desired directions.

Saying that it does not necessarily mean being able to manage them is a cautious way to put things. Strictly speaking, considering the current scientific understanding of the climate system, the body of (current and palaeoclimatic) evidence, the current trajectories of key variables, and the set of projections available, the most rigorous conclusion to be drawn is that the transformations triggered by humanity *will unfold for millennia to come.*

Naturally, the longer the destabilising human influence persists, the more these transformations will be accelerated, and the more these transformations will threaten human existence. Stopping this destabilising influence is, therefore, *of the utmost urgency*. However, if the current society is overcome, the world inherited from it will not only have a weakened capacity to support the desired qualitative abundance, but it will also have this capacity *increasingly* weakened for many centuries to come.

This is directly connected to the question of environmental struggles and the struggles of the working class. If the above reasoning is correct, it is already clear that the ultimate goal of these struggles must be the most urgent and definitive overcoming of capital. There is a reason why the term used here is "ultimate goal," not "central goal": to deliberately point out that there must be a hierarchy of goals in which the overcoming of capital is at the top. It is not just a question of suggesting a stronger emphasis or importance amidst a series of other coexisting ambitions. It is a question of suggesting that, from the point of view of our emancipatory project, the overcoming of capitalism should be a goal that subordinates all the others.

To be clear, this is not to say that the other goals and their corresponding struggles should be abandoned. What it means is that all struggles must point in the direction of this ultimate goal. Such reasoning leads to the conclusion that some struggles are dead ends and some need to be imbued with new content.

When it comes to the set of struggles that specifically concern the world of work, the reasoning is simple, although not at all trivial or intuitive. The point is that the defence of work (which broadly includes the defence of workers and the guarantee of employment) requires the class which is struggling, wanting or not, knowing or not, to defend capital itself. In this sense, struggles with this horizon are incompatible with an emancipatory and, especially, sustainable alternative.

This is no value judgement. These struggles are *legitimate.* The evils they seek to remedy are *real evils.* However, recalling the discussion above, they do not carry the potential to *spontaneously* break with order. They can, as is often claimed, be fertile ground for creating bonds of solidarity and even class consciousness. However, nothing guarantees that this class consciousness will be revolutionary. It is perfectly possible for some struggles to ferment a conservative class consciousness, as Lenin points out in his classic *What is to be Done?*.[7]

Therefore, the characteristics of the struggle very much matter. The purpose behind each one of them matters. The compatibility between means and ends matters. Some struggles point in the direction of the ultimate goal, some deviate from it, and some take us further from it. This being the case, active and constant critique is needed to distinguish between them and not take them as a whole, with a small set of abstract general criteria.[8]

Putting these intuitions into more general language: specific ends are not achievable by any means, and neither are they achievable by a disjointed set of multiple means. Even if certain ends are achievable by a certain number of alternative paths, specific ends require specific means for their realisation.

A simple illustration clarifies this point. Suppose someone wants to cut a piece of wood. That is the goal. It can be done with a chainsaw, a saw, or a sturdy knife. It can even be done with a water jet or with their hands. There are multiple means compatible with achieving the purpose. However, that does not mean it can be done by caressing the wood, strongly wishing it to split in half, or planting trees.

The same applies, even more acutely, to more complex purposes. And Marx's discussion in Chapter 11 of *Capital* reinforces this understanding.

Most of the chapter focuses on demonstrating how the productive force of labour increases when work is performed in a coordinated manner, i.e., in cooperation. However, what is crucial to the discussion of ends and means is another type of clue: the one that leads to the conclusion that some processes and achievements are simply unfeasible without a cooperative articulation of efforts on a minimum scale and scope. In other words, building on the previous discussion: if, on the one hand, it is conceivable that the achievement of some purposes is compatible with pulverised, disarticulated, non-cooperative, and incremental efforts, it is also easy to see, on the other hand, that there will be objectives for which this type of mobilisation of human action will be irremediably insufficient.

A passage from the same chapter states: "The colossal effects of simple co-operation are to be seen in the gigantic structures erected by the ancient Asians, Egyptians, Etruscans, etc."[9] This illustrates not only the superior productive force operating in combined work but also a case in which the purpose *is clearly impossible to achieve without a simultaneous and coordinated combination of efforts*. The construction of the great pyramids, for example, would be unfeasible if each small group of workers decided to adopt the procedures they found suitable without having a clear vision of the whole and how their activity would integrate that whole.

Due to the characteristics of the purpose (moving colossal blocks of stone, for example), no combination of efforts via progressive accumulation of the isolated contribution of each worker would constitute a means capable of achieving the objective. The coordination of efforts, their scale, and their application in a *determined and appropriately oriented flow* appear as *preconditions* for the achievement.

In another passage, there is a more general formulation of the same idea, which gets even closer to this issue. Marx says:

> All directly social or communal labour on a large scale requires, to a greater or lesser degree, a *directing authority* [emphasis added], in order to secure the harmonious co-operation of the activities of individuals, and to perform the general functions that have their origin in the motion of the total productive organism, as distinguished from the motion of its separate organs. A single violin player is his own conductor: an orchestra requires a separate one.[10]

In the chapter, Marx deals specifically with a moment in the development of the capitalist mode of production driven by the logic of relative surplus value.

However, it is not difficult to see that the argument is equally valid in the context of the climate crisis and the type of mobilisation needed to offer adequate responses to the challenges that will be imposed on humanity.

Speaking of large scale, the tasks imposed by climate change for human survival (and for adapting to unprecedented environmental conditions) are of such magnitude and complexity that the scientific community, not particularly known for its inclination towards subversive ideas and practices, does not even consider the possibility that pulverised and disarticulated efforts are adequate, or even acceptable. In 2014, it already warned, even if in a relatively reticent manner, that many

> adaptation and mitigation options can help address climate change, but no single option is sufficient by itself. Effective implementation depends on policies and cooperation at all scales and can be enhanced through integrated responses that link adaptation and mitigation with other societal objectives.[11]

Just four years later, the content of the recommendation is considerably more incisive. To properly understand the passage, it is important to know that overshoot indicates a trajectory that exceeds a specific target (for example, maximum temperature or maximum concentration of greenhouse gases). After the overshoot, it is assumed that it is possible to bring the level of that variable back to the goal. The excerpt states:

> Pathways limiting global warming to 1.5°C with no or limited overshoot would require rapid and far-reaching transitions in energy, land, urban and infrastructure (including transport and buildings), and industrial systems (high confidence). These systems transitions *are unprecedented in terms of scale,* [emphasis added] [...] and imply deep emissions reductions in all sectors, a wide portfolio of mitigation options and a significant upscaling of investments in those options. [...] The rates of system changes associated with limiting global warming to 1.5°C with no or limited overshoot have occurred in the past within specific sectors, technologies and spatial contexts, but there is *no documented historic precedent for their scale* [emphasis added].[12]

It is easy to see that the degree of dissonance between these recommendations and working-class mobilisation has been very high. If we are indeed the class with a vocation for emancipatory transformation, if this transformation presupposes preserving the habitability of the planet, and if this habitability is at imminent risk, then it is up to us to urgently eliminate this mismatch between what needs to be done and what has been done. The following chapters analyse alternatives that propose doing precisely that.

Notes

1 Aristotle. *Metaphysics* (2024).
2 Ibid., 88.
3 Ibid., 91.
4 Aristotle, *Metafísica* (2002). Translator's note: The English translation of this passage differed significantly from the Portuguese translation used by the author to develop his reasoning. Therefore, we have chosen to translate the passage as found in the Portuguese version of the text.
5 Lukács, *Para uma ontologia do ser social II [Ontology of Social Being], vol. 2* (2013), 55.
6 White, Gareau, & Rudy, "Ecosocialisms, past, present and future" (2017).
7 Lenin, *Collected works*, vol. 5 (1977).
8 This kind of appropriation as a whole is found, for example, in the following passage by Burkett:

> Once one drops the strictly industrialist conception of production and class struggle, it becomes clear that all self-activated popular struggles over existing social conditions, and for new conditions amenable to a less restricted human development, are integral to the general movement toward an associated form of production conceived as a new union of the producers and the social conditions of their production.

Burkett, *Marx and nature* (1999), 213.
9 Marx, *Capital,* vol. 1 (1982), 451.
10 Marx, *Capital,* vol. 1 (1982), 448–449.
11 IPCC, *Climate change 2014* (2014), 26.
12 IPCC, *Global warming of 1.5°C* (2018), 21.

References

Aristóteles. (2002). *Metafísica.* São Paulo: Edições Loyola.
Aristotle. (2024). *Metaphysics.* E-bookarama.
Burkett, P. (1999). *Marx and nature: a red and green perspective.* New York: St. Martin's Press.
IPCC. (2014). *Climate change 2014: synthesis report.* Geneva: IPCC.
IPCC. (2018). *Global warming of 1.5°C: an IPCC special report on the impacts of global warming of 1.5°C above pre-industrial levels.* Geneva: IPCC.
Lenin, V. I. (1977). *Collected works* (Vol. 5) Moscow: Progress Publishers.
Lukács, G. (2013). *Para uma ontologia do ser social II.* São Paulo: Boitempo.
Marx, K. (1982). *Capital* (Vol. 1). London: Penguin Books.
White, D. F., Gareau, B. J., & Rudy, A. P. (2017). Ecosocialisms, past, present and future: from the Metabolic Rift to a reconstructive, dynamic and Hybrid Ecosocialism. *Capitalism, Nature, Socialism, 28*(2), 22–40.

13 Green New Deal and the enticing legislative route

In the last week of November 2019, the World Meteorological Organization released its 2018 consolidated data on CO_2 concentration levels in the atmosphere. According to its report on greenhouse gases, this concentration reached levels "observed" for the last time 3 million years ago.[1] To provide a clear idea of the magnitude of this time scale, current estimates indicate the most distant ancestors of the genus Homo date back this far. Modern man, on the other hand, only appeared 200–300 thousand years ago.

As discussed in Chapter 11, stabilising this process of chemical transformation of the atmosphere would require sharp and sustained reductions in annual emissions of various gases. Another report, released almost simultaneously by the United Nations Environment Programme (UNEP), warns that global greenhouse gas emissions per year would have to be reduced by 7.6% a year for the next ten years *just to meet the goals set by the Paris Agreement.*[2] However, not even the containment of the growth trajectory had been achieved until 2020.

Since the Covid-19 pandemic began to halt a series of important economic activities, news and assessments about possible ecological impacts started to circulate. The first effects were seen in the superficial regeneration of the urban landscape, with the sky becoming bluer, the air cleaner, the water in canals clearer, animals roaming the streets, and more. Then, there were projections of a reduction in greenhouse gas emissions in 2020. Initially, a reduction of 1.5%, then corrected to 4%, and finally to 8%.[3] Today, the IPCC estimates that the reduction was 7%.

It is easy to see that this one-off reduction in the level of global emissions comes at an immense human, social, and economic cost resulting from the forced interruption of the flows of people and goods around the world. What requires more careful demonstration is the fact that it is impossible to achieve the reductions recommended by the Intergovernmental Panel on Climate Change (IPCC) without sacrifices of a similar magnitude, as discussed in Chapter 11. If the emissions reduction trajectory proposed by the IPCC is correct, annual reductions of more than 7% for 30 consecutive years are necessary, as well as a

DOI: 10.4324/9781003512516-16

carbon capture technological apparatus that does not even exist yet implemented on an unprecedented scale, scope, and speed.

However, what is usually forgotten when the virtuous characteristics of new technologies are mentioned is that all of them, without any significant exception, are protected by patents. By way of illustration, at the end of 2019, CNN published online the story of a start-up that had achieved an extraordinary energy breakthrough using solar technology controlled by artificial intelligence.[4] They said this would finally make it possible to decarbonise important heavy industries, responsible for a significant fraction of global CO_2 emissions. Therefore, CNN itself declared in bold letters: this is a technology to "solve the climate crisis." In the body of the text, however, it is casually mentioned that the technology is patented. This is the heart of the matter.

Suppose the new technology does indeed have all the alleged potential to eliminate greenhouse gas emissions in very polluting industries. All over the world, scientists, activists, and even politicians are warning that we are living through a climate emergency that requires *immediate* and *combined* global action. Humankind does not have 20 or 30 years to wait for the normal course of innovation to spread. Therefore, if this new technology does indeed have the alleged material impact, the next and irrevocable step should be its *accelerated universalisation*. In other words, whatever the potential benefit, it will only be realised if this technology spreads rapidly throughout the world and to all possible uses.[5] But it has been patented, which implies legal and economic restrictions on its use.

The patenting of supposedly life-saving innovations does not *simply* reveal the pathological greed of mega-entrepreneurs, a misdirected political will, or a poorly designed industrial policy. Rather, this practice is a symptom of something deeper, of the very logic of how capitalist society works. In capitalist society, useful objects are only produced if they pass through the filter of profitability. In other words, only those objects which can be produced with profit will be. Utility and human needs are absolutely secondary.[6] Even if the need is, for example, to "save the world" or "solve the climate crisis."

Some of the technologies and ways of life that would allow humanity to survive a rapidly changing world already exist and are known. But they are still bound by structures whose function is only to guarantee their economic viability (i.e., profitability), restricting their use. Its urgent universalisation, therefore, requires the removal of these ties.

Aronoff and her co-authors do not fail to explicitly recognise this requirement, albeit in a somewhat moderate way: "we want to radically loosen patents to speed global cooperation in clean tech, making the best tools available to all countries."[7] It should come as no surprise that this requirement is completely absent from the US Green New Deal[8] and the European Green Deal.[9]

However, we are not only faced with the challenge of finding and generalising new technologies but also (and especially) the imperative of eliminating old ones. The largest share of emissions, just over a third of the total, comes from energy production.[10] In the Green New Deal literature, we find clear recognition that a radical decarbonisation of the world's primary energy supply is, therefore, unavoidable.

Proven reserves of oil, coal, and natural gas must remain untapped. Simply encouraging renewable sources is nowhere near enough. The boost to the proliferation of these sources in recent decades has been extraordinary, and yet it has not been able to displace fossil fuels from the world energy matrix. However, there is no clear recognition that, much more than a purely technical issue, what is involved here is a fierce struggle against one of the most powerful sectors of contemporary capitalism.

To further support this insight, it is worth mentioning that the untapped reserves of fossil fuels circulate as trillions of dollars in financial assets on the futures markets, and the corporations themselves have their value partly determined by what *is still underground.* Challenging fossil fuels, therefore, does not only involve fighting big oil and related sectors. It involves confronting the entire financial sector head-on, all its direct players, and all the professional politicians on its payroll.

In this case, it is also worth emphasising that it is the hypertrophy of the financial sphere in contemporary capitalism that determines the relatively solid hegemony of the neoliberal agenda, not the other way around.[11] Therefore, it is disconcerting that a Marxist like John Bellamy Foster believes in overcoming fossil fuels without overcoming capitalism, postponing the revolutionary transition to an indefinite "long term." In a 2019 interview, the author says:

> In the long run, of course, we have to have a full ecological and social revolution, transcending existing capitalist relations of production. But right now, we are in an emergency situation, and the first priority is eliminating fossil fuels, which entails the destruction of what is called fossil capital.[12]

Returning to the list of challenges that usually appear in Green New Deal literature, agriculture and land use in general, which account for just under a quarter of total emissions, also require a comprehensive response. The strong mitigation needed in this area implies confronting the interests of agribusiness, which necessarily includes confronting profit and private land ownership. Industrial agriculture, monoculture, deforestation, and the massive use of nitrogen fertilisers (which emit nitrous oxide) make up a destructive system that must undergo a complete overhaul under public control and informed by the best agroecological practices. Once again, this is not an enemy of modest powers. And, once again, little is said about how to confront it and, above all, defeat it, other than the traditional legislative and regulatory route.

Last but not least, there is transportation. This is an emblematic case because many of the most apparent effects mentioned at the beginning of the text are related to this sector. The reduced circulation of cars in large cities under social distancing measures or mandatory lockdowns led to rapid improvements in urban air quality. The significant contraction in tourism and the closing of borders in several countries have also had an impact on air traffic, one of the most carbon-intensive human activities.

But these disarticulated, improvised, and temporary retractions should not serve as a model. It is impossible to imagine that transport, which is responsible for around 14% of total emissions, will show the necessary reductions without certain modes being overcome, for example, the individual private car and the plane, and without the private logic that operates public transport being eradicated once and for all. As Aronoff et al. point out, "the goal for cities and suburbs is a public transportation system that is free, and so good that most people will support major restrictions on private cars (and ultimately a ban)."[13]

It is impossible to challenge the relevance of this goal. However, without an emphatic affirmation of the need to overcome capital, it loses substance. If this goal has not been achieved even by those central countries that have built the most extensive welfare states, what could lead to believing that it will now be achieved globally, precisely at the current stage when the dynamic vigour of capitalist society is showing increasing signs of exhaustion and instability?[14]

Achieving any one of these results on a national level would be a monumental task. However, it is essential that *all of them* are achieved *and globally*. There is no point in a country achieving significant or complete decarbonisation at the expense of "exporting emissions." Geopolitical disputes over control of technologies, mineral resources, and borders are absolutely incompatible with scenarios that guarantee at least a minimum chance of preserving our species. Even so, as Aronoff, Battistoni, Cohen, and Riofrancos state, the major economic and military powers of the planet are already moving to secure privileged access to supply chains that will be decisive in the near future, such as the lithium supply chain.

This very brief overview of a few central points is enough to conclude that it is essential to confront some of the most powerful pillars of capitalism today. And to be clear: to avoid our own demise, defeat or partial victories are not an option.

Given all these challenges, it is disturbing that the big bet of today's environmentalism is the so-called Green New Deal. Right away, it is worth pointing out the completely inappropriate choice of drawing an intentional analogy with economic policies that saved capitalism in the 1930s, especially considering what Naomi Klein herself acknowledges:

> the New Deal was adopted by Roosevelt at a time of such progressive and left militancy that its programs (which seem radical by today's standards) appeared at the time to be the only way to hold back a full-scale revolution.[15]

The question remains: if we need an effectively revolutionary transformation, *is it not a waste of precious time* to mobilise efforts for an agenda that is restricted *a priori* to the limits of the reproductive parameters of capitalist society?

Furthermore, the so-called Green New Deal suffers from a more debilitating problem: it aims to advance this whole agenda of profound transformations in the parliamentary arena, using laws. The rather desiccated bill presented in the US Congress (and not even endorsed by the majority of Democrats) and the even more backward version approved in the European Parliament clearly exhibit the long-known limits of this strategy.

Given the blatant frustration of the most important ambitions present in the debate and even the frustration of the most modest goals that survived transposition into the institutional arena, it is not uncommon for these results to be attributed to an ideological surrender to business as usual, a lack of political will, or a lack of capacity to achieve effective international articulation. All these factors undeniably act as barriers to progress on the climate agenda. However, three considerations must be made.

First of all, if business as usual is the set of practices that seek the maximum possible growth of capital at the lowest possible private cost and with absolute indifference to any social cost, the word *capitalism* can easily replace this term. It is not about a version of capitalism going wrong or deviating from its otherwise virtuous course. It is just capitalism stripped down to its most fundamental essence. Therefore, in its spontaneous unfolding, free from external pressures and State regulations, this nature of capitalism fully reveals itself.

Second, this apparent political paralysis should be understood less in moral/subjective terms and more in structural/objective terms. Indeed, there is usually room for external pressure and State regulation. In other words, there can be some progress towards forcing large corporations to incorporate part of their social and ecological operational costs as private costs. This may limit destructive practices, block anti-ecological activities, and even force transitions to more ecologically sustainable practices and activities. This space for small victories, however, is not unlimited. In fact, it is pretty narrow since no measure that threatens the foundations of profitability and competitiveness is ever tolerated. Therefore, it is not just a question of lack of political will to do what is necessary. The point is that the requirements capital and its representatives cannot give up structurally limit that political will.

Third, to illustrate this point, the last few decades have been characterised by an unprecedented number of climate policies.[16] Even if the most apparent result is a certain paralysis in the major international coordination spheres, there have been very significant results. For example, primary energy production from so-called renewable sources grew by 157% between 1973 and 2017. Taking 1992 as a base, there is an expansion of 66.6%. Moreover, the biggest increases in these 45 years, in absolute terms, have been concentrated in the last 26 years, strongly motivated by the First Earth Summit (Eco-92, in Rio de Janeiro) and all the subsequent climate and environmental conventions. However, this substantial result has not been able to displace fossil fuels from the world energy matrix, which in 1990 accounted for 81.19% of all energy produced and in 2017 accounted for 81.21%.[17]

Apparently, this is a paradox. If important results have been achieved, how to explain the generalised (and faster than expected) deterioration of the main climate variables and the persistent failure to achieve the most fundamental goals? Curiously, the explanation is simple, especially after the entire theoretical basis presented in the first part of this book.

The desired result of all State regulation of certain activities, all taxation imposed on various pollutants, and all incentives for alternative forms of energy, etc., is the moderation of our material impact on the planet. When these policies force economic agents to take the ecological impacts of their practices as a private cost, the expected result is that these practices are reduced or eliminated, especially by reducing resource consumption and the generation of polluting waste. The problem, however, is that although the conditions for this moderation of our impact are created all the time, *they are also blocked/neutralised all the time*, as Chapter 4 explained in detail.

At the beginning of this chapter, it was said that the current state of atmospheric chemistry is unprecedented for our species and even for our most distant ancestors. The last time it was in the same condition, it is estimated that the average temperature of the planet was 2–3°C higher than today and that the sea level was 10–20 metres higher than today. It is not at all extravagant to conclude that we *already live* in a world unknown to the human species. Furthermore, this world is undergoing an abrupt process of transformation. Many of these transformations could even, in the terribly near future, jeopardise Earth's ability to sustain life.

The burgeoning Green New Deal literature recognises, to a large extent, the dramatic contours of this crisis. However, having mapped out with relative precision the magnitude of the transformations that must be achieved to avoid a collapse of civilisation, it proposes challenging all the pillars of modern capitalist society *in the sphere of institutional politics*, precisely one of the main arenas in which conflicts find the most appropriate resolutions for the reproduction *of capitalist society*. Moreover, it is precisely the arena in which resolutions that are incompatible with the logic of capital are buried. In other words, it is an agenda that proposes to overcome capitalism without challenging it.

It is true that there is a diffuse and insistent call for popular pressure that would somehow be able to redefine the range of possibilities within institutional politics. Indeed, depending on the level of these pressures and the new balances of power established, it is possible to obtain certain concessions from capital. However, as Gramsci says, "there is no doubt that such sacrifices and compromise cannot involve the essential."[18] And when it comes to the ecological issues dealt with here, the essentials can easily be understood as profit, private property, and growth. That is why these challenges demand a popular mobilisation with truly insurrectionary energy capable of taking over and establishing collective control of the directions that need to be taken. If this mobilisation is not ready, our most immediate task is to create it. Overcoming capitalism is the minimum programme for confronting the climate crisis that lies ahead.[19]

Notes

1 WMO, *Greenhouse gas bulletin* (2019).
2 See UNEP, *Emissions gap report 2019* (2019). The IPCC warns that full compliance with this agreement would still be insufficient to limit global warming to 1.5°C above pre-industrial temperatures. IPCC, *Global warming of 1.5°C* (2018).
3 See IEA, *Global energy review 2020* (2020).
4 Egan, Secretive energy startup backed by Bill Gates achieves solar breakthrough (2019). CNN Business. See https://edition.cnn.com/2019/11/19/business/heliogen-solar-energy-bill-gates/index.html
5 There are also all kinds of technologies and technological advances that do not even have the potential to be ecologically beneficial. Sá Barreto, "Marx contra o otimismo tecnológico: economia 'imaterial' desmistificada e desdobramentos para as questões ambientais" ["Marx against technological optimism: demystifying the 'immaterial' economy and the consequences for environmental issues"] (2016).
6 Mészáros, *Beyond capital* (1995).
7 Aronoff, Battistoni, Cohen, & Riofrancos, *A planet to win* (2019). No page numbers are provided as the quote was taken from an electronic book.
8 Ocasio-Cortez, *Recognizing the duty* (2019).
9 European Commission, *The European Green Deal* (2019).
10 United States Environmental Protection Agency, Global Greenhouse Gas Overview (2014). See www.epa.gov/ghgemissions/global-greenhouse-gas-emissions-data
11 Sá Barreto, "Restauração neoliberal e o esgotamento histórico de formas consagradas de resistência" ["Neoliberal restoration and the historical exhaustion of established forms of resistance"] (2019).
12 Foster, "Interview" (2019).
13 Aronoff, Battistoni, Cohen, & Riofrancos, *A planet to win* (2019). No page numbers are provided as the quote was taken from an electronic book.
14 Sá Barreto, "Restauração neoliberal e o esgotamento histórico de formas consagradas de resistência" ["Neoliberal restoration and the historical exhaustion of established forms of resistance"] (2019).
15 Klein, *On fire* (2019). No page numbers are provided as the quote was taken from an electronic book
16 Sá Barreto, *O capital na estufa: para a crítica da economia das mudanças climáticas* [*Capital in the greenhouse: a critique of the economics of climate change*] (2018).
17 International Energy Agency, Energy Statistics Data Browser (2014). See www.iea.org/data-and-statistics?country=WORLD&fuel=Energy%20supply&indicator=Total%20primary%20energy%20supply%20(TPES)%20by%20source
18 Gramsci, *Cadernos do cárcere* [*Prison notebooks*], vol. 3 (2014), 49.
19 This somewhat controversial claim will be covered in more detail in Chapter 16.

References

Aronoff, K., Battistoni, A., Cohen, D., & Riofrancos, T. (2019). *A planet to win: why we need a Green New Deal.* London: Verso.

European Commission. (2019). *The European Green Deal.* Brussels: EC.

Foster, J. B. (2019). *Interview: John Bellamy Foster on the 'Green New Deal'.* Retrieved from Climate and capitalism: https://climateandcapitalism.com/2019/02/12/john-bellamy-foster-on-the-green-new-deal/

Gramsci, A. (2014). *Cadernos do cárcere* (Vol. 3). Rio de Janeiro: Civilização Brasileira.

International Energy Agency. (2020). *Global energy review 2020: The impacts of the Covid-19 crisis on global energy demand and CO2 emissions.* Paris: IEA.

IPCC. (2018). *Global warming of 1.5ºC: an IPCC special report on the impacts of global warming of 1.5°C above pre-industrial levels.* Geneva: IPCC.

Klein, N. (2019). *On fire: the burning case for a Green New Deal.* Toronto: Penguin Random House.

Mészáros, I. (1995). *Beyond capital.* London: Merlin Press.

Ocasio-Cortez, A. (2019). *H.Res.109 – Recognizing the duty of the Federal government to create a Green New Deal.* Washington, DC. Retrieved from www.congress.gov/bill/116th-congress/house-resolution/109/text

Sá Barreto, E. (2016). Marx contra o otimismo tecnológico: economia "imaterial" desmistificada e desdobramentos para as questões ambientais. *Nova Economia, 26*(1), 97–122.

Sá Barreto, E. (2018). *O capital na estufa: para a crítica da economia das mudanças climáticas.* Rio de Janeiro: Consequência.

Sá Barreto, E. (2019). Restauração neoliberal e o esgotamento histórico de formas consagradas de resistência. *Revista da Sociedade Brasileira de Economia Política, 53,* 118–146.

UNEP. (2019). *Emissions gap report 2019.* Nairobi: UNEP.

WMO. (2019). *Greenhouse gas bulletin: the state of greenhouse gases in the atmosphere based on global observations through 2018.* Geneva: WMO.

14 An ecosocialist critique of developmentalism and the energy transition

By reading the previous chapter on the Green New Deal and texts by ecosocialists on the subject of the energy transition, it is possible to notice some important similarities between the propositions that are being elaborated and developed. Precisely because of this superficial similarity, a warning must be immediately made. In general, ecosocialists approach these issues from a Marxist perspective, unlike many of the leading formulators of the Green New Deal agenda. As such, they approach this debate seeking to make it as tense as possible, disputing its direction and imaginary. This chapter is a critique of the way this is done. However, it deals with conceptions which, unlike those discussed in Chapter 13, have the struggle for an ecosocialist transition as their explicit horizon.[1]

This chapter is organised around three main topics. First, the sensitive issue of inequality, discussed in the light of the centre/periphery structure, brings with it the issue of developmentalism, with its ideology of progress, industrialism, growth, and more. Second, the alternative of qualitative growth, which is present both in ecosocialist interventions and in the Green New Deal literature, and even in Ecological Economics, in different versions. This topic was part of the discussion of qualitative abundance in Chapters 8, 9, and 12. Now, the problem will be approached from a different angle. Third, and most importantly, the defence of an energy transition, which will require the consideration of a *broader* notion of green capitalism.

The tension between the centre and the periphery of the capitalist system emerges every time ecological reflection comes to the conclusion that the scale of human activities, demand, and material impact needs to recede quickly and intensely. What is usually claimed is that the idea of degrowth (of production and consumption) is absurd and has a certain imperialist accent, as it would trap various nations of the world in a situation of misery.

There are many on the left who, to this day, suffer from a kind of nostalgia for the "Golden Age" of capitalism, assuming that a virtuous development process

DOI: 10.4324/9781003512516-17

must first solve serious social problems so that, only then, it will be possible to tackle the ecological challenges. The understanding of development varies greatly, but traditions leaning more to the left agree about a few of its supposed features.

The first common feature is the idea of progress, i.e., the notion that development is always a transition from a worse stage to a better one. This positive interpretation of development is incompatible with the Marxist perspective. According to Marx, development is the process of complexification of a being; of making its essence explicit and manifest. Applied to capitalist society, this understanding of what development is involves making the essence of capital explicit in an increasingly comprehensive and profound way.[2] In other words, development would be the process in which society, and the world in general, becomes more and more the reality *posited by capital.*

In this world built in the image and likeness of capital, some regions enjoy better conditions than others, and some countries have risen from relative penury to a more comfortable material and social situation. This does not happen by chance or deviating from a normal virtuous course. The structure that forms and consolidates a hierarchy between centre and periphery *is the very expression of development in capitalism*, which leads to the second common feature.

The second feature is the means to move up this hierarchy. There are two main ways to break away from the peripheral condition. Either the country climbs the world economic hierarchy, gaining more advantageous positions in the international division of labour, or it *breaks* with this hierarchy. The second path, which corresponds to a necessarily revolutionary transition, is generally left to an indeterminate future. As far as the first path is concerned, since it is a process by which the country must accumulate the capacity to impose itself internationally, at least in economic terms, development strategies up to now have always involved at least an attempt at rapid industrialisation.

The third common feature, which may be the basis of the positive assessment that developmentalism still has on the left, is that development is seen as a process that enables a country to build a social protection network based on public welfare, universal and quality health and education, among other advantages. It is not hard to see that no country with a weak economy would be able to sustain such a welfare State. Whichever welfare State model one takes as a reference, it is safe to say that they all require a healthy national economy.[3] In capitalism, a healthy national economy is an economy that grows. If, on the one hand, developmentalism has close ties to an industrialist ideology, on the other it is absolutely inseparable from the obsession with growth.

Finally, re-establishing contact with the ecological discussion, it is usually said that, considering the world's situation as it is, simply admitting the need to degrow would cause rich countries to use their geopolitical and economic power to impose degrowth on poor countries. That, of course, is true. Indeed, it is quite implausible to imagine that the world's greatest powers would voluntarily give up a substantial part of their material affluence. However, the rest of this reasoning is a constellation of errors.

First, it is a historical error to assume that the path to development in capitalism is open to any nation that manages to put together a suitable project, sufficient political will, and sufficient resources. This overlooks the fact that only a small number of countries have managed to reach a stage of economic development that would make a reasonably comprehensive and universal social protection network feasible. Even among these countries, there are those in which significant fractions of the population live in material hardship. The history of the 20th century shows that such development is an *extremely rare* experience. When economic development is placed as a precondition for incisive ecological mobilisation, what we are doing, strictly speaking, is halting the tackling of climate issues from the outset.

Second, there is a political error. Developmentalists correctly recognise that there is no predisposition on the part of the global powers to give up their expensive way of life. On the other hand, they seem to assume that, by some magical spontaneous mechanism, these same powers would bovinely tolerate a mass rise of all the other nations on the globe and all the disruptions this would bring to the international division of labour and the hierarchies of power.

Third, there is a quantitative error. According to data presented by Professor Luiz Marques in his monumental book *Capitalism and Environmental Collapse*, the poorest 3 billion people in humanity emit only 7% of the annual CO_2. At the same time, only 8% of the world's population owns more than 80% of global wealth.[4] In addition, as we already know, recent studies recommend that global emissions be halved in the short term and reduced to zero in the medium term. These three pieces of information reinforce what was foreseen in the previous paragraph. If there is no reason to harbour illusions of voluntary and generous degrowth in the countries of the centre, there is no reason to believe that they would "authorise" the development of peripheral countries (however the term is understood). On the contrary, it is more reasonable to imagine that the great powers are directly interested in the complete extermination of populations at the margins. However, as the scale of the poorest half emissions (7%) indicates, not even that would be enough.

Finally, and most seriously, there is a systemic error. The paranoid denial of the existence of climate change usually inspires scorn and derision from progressive individuals and organisations, because it is blatantly ignorant and anti-scientific. However, this arrogant attitude completely changes when science starts to indicate that the changes are already irreversible and abrupt, that humans are, therefore, in the midst of a climate crisis, and that a climate collapse is a very tangible (and even probable) possibility. Only by completely ignoring the recent advances in understanding the changes taking place is it possible to reject the idea that humanity is facing a climate emergency.

Therefore, even supposing economic development is a universally desirable and achievable goal within the economic and geopolitical frameworks of capitalism, the most recent scientific production affirms that there is no longer time for it. The choice is no longer between wealth or poverty, affluence or penury, comfort or misery. At the moment, the real choice humankind faces is between the survival of the species or its virtual elimination from the face of the Earth.

It is worth quickly recalling the structure of this chapter. It is organised around three main topics. The first one discussed the centre/periphery structure and developmentalism. The following topic concerns the proposition of qualitative growth as an alternative to productivist or merely quantitative growth.

In Löwy, the treatment of the issue is less than rigorous. Some passages are emblematic of how the problem is simply brushed aside through a rhetorical trick, such as the one below:

> There would be no need – as some puritanical and ascetic ecologists seem to believe – to lower the standard of living of European or North American populations in absolute terms. It would simply be necessary for these populations to get rid of useless products, those that satisfy no real need and whose obsessive consumption is sustained by the capitalist system.[5]

The first thing to address in this passage is its very form. Being rigorous with what is written, Löwy is basically saying that there is no need to lower the standard of living in absolute terms, it would only be necessary to lower the standard of living in absolute terms.

The reasoning only stands by adhering to an ahistorical, abstract, ideal notion of what a real need is. Löwy, as a Marxist, should not do that. In the second paragraph of the first chapter of Book I of *Capital*, Marx makes it very clear that he does not understand *need* in this rigid way. He says: "The commodity is, first of all, an external object, a thing which through its qualities satisfies human needs of whatever kind. The nature of needs, whether they arise, for example, from the stomach, or the imagination, makes no difference."[6] In a footnote, he quotes Nicholas Barbon: "Desire implies want; it is the appetite of the mind, and as natural as hunger to the body (...) The greatest number (of things) have their value from supplying the wants of the mind."[7]

Evidently, none of this implies that the need for trinkets is legitimate or should be preserved. The point is that the need for trinkets is a *real* need in this society. As such, Löwy cannot simply assume that ridding the European or North American populations of their useless products is an easy or even simple task. Nor can he assume that it is a task of mere persuasion or legislation and regulation. He rightly recognises that the consumption of these useless products is fuelled by capital. However, he does not seem to realise that capital does it out of necessity. Chapters 5 and 6 on consumption explain this necessity in further detail. Wasteful consumption is a *pillar* of capitalism and capital cannot do without it.

Non-Marxist versions of reasoning similar to Löwy's say that growth, even merely quantitative growth, can continue by promoting the growth of green sectors and activities and imposing the decline (and eventual suppression) of dirty sectors and activities. This similarity is not a mere detail. Strictly speaking, Löwy is at least flirting with a conception of green capitalism in this passage. The topic will be examined in detail in the discussion of energy transition.

Fernandes approaches the question more interestingly. She says: "Ecosocialism breaks apart the meaning of development in order to rid it of its capitalist and colonial facets and enrich it with qualitative – rather than merely quantitative – notions of a good life."[8] Fernandes is alluding to the need to dispute the imaginary around the notions of development and growth.

At the risk of oversimplifying a difficult and complex subject, some things must be considered. There is always an immense number and variety of ideas and conceptions in circulation. Some are reconcilable, others irreconcilable. Some better apprehend how things factually work, others do it incorrectly or inconsistently. Some will be truer than others. This book has covered some of these multiple ideas. Productivist ideology circulates, as do concepts related to a simple life. The obsession with growth circulates, as do propositions relating to qualitative abundance. Wasteful ethics circulate, as do ideas of conscious consumption.

Indeed, it is not hard to see that although all these ideas circulate, some remain marginal while others are persistent and dominant. When disputing the imaginary, one needs to be careful not to give in to the temptation to think that the hegemony of certain ideas depends on their quality or, worse, to think that it would be possible to elevate another set of values to the dominant position through some process of persuasion. Marx knew that is not how things work. In a famous passage from *The German Ideology*, he and Engels state that "The ideas of the ruling class are in every epoch the ruling ideas: i.e., the class which is the ruling material force of society is at the same time its ruling intellectual force."[9]

In the light of this insight, it becomes clear that the current notion of development is not the dominant notion by chance. It is the dominant vision because it mirrors the real meaning that development has in this society, which is the accumulation of capital and the colonisation of every thread of the social fabric by the logic of capital. In other words, it is the vision that is effectively compatible with the reproduction of this society.

The great difficulty, therefore, is that disputing the imaginary is *necessary but insufficient*. To overturn the predominance of productivist, expansionist, wasteful, utilitarian ideas, it is unavoidable to challenge the very society that establishes and requires the widespread validity of these ideas. In other words, if it is the very metabolism of capital that requires individuals to behave productively, utilitarianly, regardless of the conflicting ideas and values that these individuals may carry within themselves, then the possibility of the free and widespread exercise of a non-destructive ethics only opens up with the destruction of the metabolism of capital.

The third and final point of the chapter is the energy transition, a priority task of the so-called Green New Deal. Likewise, ecosocialists place great emphasis on this objective. In both cases, the emphasis on the energy transition stems from the urgent need to decarbonise how we live. It is important to draw a line between two concepts usually taken as roughly equivalent: energy transition and decarbonisation. The significance of this difference will become clearer as the discussion unfolds.

Fernandes is one of the ecosocialists to address the issue. Her approach may be analysed in three short passages that seem to be representative of her understanding. First:

> Ecosocialists are well aware that only a revolutionary path can take us beyond the capitalist system. But they also understand that other clusters of change and reforms must garner support in their radical forms before a pre-revolutionary scenario appears on the horizon.[10]

When this reasoning is placed within the framework of the energy debate, it reveals an understanding that is not just Fernandes'. This understanding is that an energy transition would preserve the material conditions for the moment of the revolution. The necessary energy transition is understood as a cluster of changes, a reform within the system. There is a logical assumption, therefore, that the energy transition is a goal that can still be achieved under capitalism.

The second passage presents this image in all its colours: "A decarbonized mode of production is necessary to ensure that when workers are ready to overthrow capitalist structures, there is still a healthy planet on which to build socialism."[11] In this passage, the idea that sums up the energy transition proposal is expressed in full. The objective (which is intermediate, at least for ecosocialists) is decarbonising capitalism while waiting for the revolution.

Finally, there is the following: "Decarbonization is both an immediate necessity and a material prerequisite for any transitional program and the very prospect of organizing to abolish private property."[12] This formulation is the product of a dead end that Fernandes and many others set up for themselves, usually right at the beginning of their reflections. When they say there is no possibility of a revolutionary process on the horizon, all the alternatives that would only be viable in such a process are halted from the start. At the same time, as Fernandes is rigorous in working with the best available science, there is also the explicit recognition that decarbonisation is urgent and unavoidable. However, since the revolutionary path has been removed from consideration, there is only the alternative of conceiving forms of decarbonisation *within capitalism.* Since capitalism cannot reproduce itself on a decreasing scale, the only conceivable form of decarbonisation is energy transition.

In other words, if decarbonisation is necessary and supposedly the current order cannot be overthrown, we must decarbonise under capitalism. Furthermore, if it is not permissible to shrink the *scale* of energy consumption under capitalism, we must cleanse energy consumption of its carbon content. In other words, we must carry out the energy transition.

Therefore, a deeper analysis reveals that the energy transition agenda has a "touch" of green capitalism. If it is possible to clean up the energy matrix to the point of achieving the necessary level of decarbonisation, then, at least from an energy point of view, capitalism can be clean. This result has to be rejected. And this is not an *a priori* rejection.

The previous chapters demonstrated that the ecological critique that Marxism can direct at capitalism maintains very solidly that this system *cannot be green.* Thus, even in the face of an obstacle that seems insurmountable, it is of utmost importance to remain consistent with this demonstration and conclude that any development alternative with green contours requires a revolutionary process.

It is understandable that given the urgency and magnitude of the challenge, practical reflection may slide back to the list of initiatives that are already more or less known and are already in motion to some extent.[13] However, it needs to be clear that not only market-based solutions are "green capitalism". All the solutions that claim to be solutions without triggering a historical process of breaking with capital share with green capitalism an overvaluation of the plasticity of this system. In other words, they all convey, even if they do not want to, the idea that a better capitalism than the current one is possible. In the specific case of the energy transition, a decarbonised capitalism or one in the process of decarbonisation.

The whole theoretical discussion in the first part of this book leads to the understanding that the initiatives in search of this transition cannot even contain the destructive impetus of human activity as long as they are subject to the social metabolism corresponding to capital.

This inability to move forward can easily be illustrated by the recent history of the global quest for decarbonisation. As mentioned in the previous chapter, between 1992 and 2017, the global primary supply of energy from renewable sources expanded by just under 67%. This is a significant result. However, it has not been able to effectively replace fossil fuel consumption. In other words, it has not been able to trigger an effective energy transition.

Once again, the reasons are evident. In the same period, the primary energy supply from natural gas, coal, and oil increased by 80%, 78%, and 38%, respectively. In terms of absolute increases in primary energy supply, renewable sources are in fourth place, with an increase around ten times lower than that of third place, oil. In first and second place, respectively, are coal and natural gas.[14]

None of this is by chance. This insane and feverish mode of production called capitalism is absolutely dependent on cheap and abundant energy. Without it, the gears of production for production's sake jam, and the dynamics of accumulation chokes. The assumption that it is possible to take steps towards an energy transition under capitalism loses sight of this fact. Its proponents do not even realise, for example, that all of today's renewable energy supply, including that from the hundreds of nuclear power plants scattered around the world, would only be enough to meet the energy demand of the early 1950s.

Indeed, despite the digging done here, which led to the conclusion that the idea of energy transition has a similar content to other green capitalism proposals, it must be underlined that this proposal, when incensed by ecosocialists, has at least a questioning form. For example, Fernandes says: "The logic must be to decarbonize fast, with a focus on the public system, fighting privatisation at all costs, and strengthening popular movements and organisations."[15] In this sense, if this agenda challenges the logic of capital, then it is part of a revolution (if it intends to successfully challenge that logic), and not part of a process restricted to the reproductive parameters of capitalism. Otherwise, this agenda only challenges the *subjectivity* of capital, its typical rationality, being achievable only to the extent that it is stripped of its disruptive potential. However, if stripped of its disruptive potential, it will no longer be able to deliver the results it set out to deliver, regardless of its initial goals and ambitions.

Disputing this subjectivity, disputing the imaginary, is important. But the kind of framing given to the issue prevents us from asking ourselves a crucial question. Given that decarbonisation is necessary and urgent, given that it is not obtainable under capitalism, and given that, even in a revolutionary scenario, there would be technical restrictions that would make decarbonisation much slower than necessary, what tends to happen? That is, given that decarbonisation will probably not happen in time to prevent a catastrophe, what will be the more or less predictable characteristics of that catastrophe? How to prepare and organise responses to it? These are the questions discussed in the next chapter.

Notes

1 See Fernandes, "Ecosocialism from the margins" (2020a); Fernandes, "Una visión ecosocialista para la descarbonización en América Latina" ["An ecosocialist vision for decarbonization in Latin America."] (2020b); Löwy, *O que é o ecossocialismo?* [What is ecosocialism?] (2014).

2 Bonente, *Desenvolvimento em Marx e na teoria econômica: por uma crítica negativa do desenvolvimento capitalista* [Development in Marx and in economic theory: claiming a negative critique of capitalist development] (2016).

3 Esping-Andersen, *Welfare capitalism* (1990).

4 Marques, *Capitalism and Environmental Collapse* (2020).
5 Löwy, *O que é o ecossocialismo?[What is ecosocialism?]*(2014), 91.
6 Marx, *Capital,* vol. 1 (1982), 125.
7 Ibid.
8 Fernandes, "Ecosocialism from the margins" (2020a), 139.
9 Marx & Engels, *Collected works*, vol. 5 (1976), 59.
10 Fernandes, "Ecosocialism from the margins" (2020a), 137.
11 Ibid., 142.
12 Ibid., 142.
13 It is worth remembering that Marx himself, in another context, pointed out something similar: "And just when they seem engaged in revolutionising themselves and things, in creating something that has never yet existed, precisely in such periods of revolutionary crisis they anxiously conjure up the spirits of the past to their service and borrow from them names, battle-cries [...] in order to present the new scene of world history in [...] this borrowed language." Marx & Engels, *Collected works*, vol. 11 (1980). The sour tone of this passage would be much more rightly directed at proponents of the Green New Deal. However, I chose to bring it up here to emphasise that not even we Marxists are immune to incorporating agendas that bind us to the reproduction of the status quo under the illusion that we are challenging it.
14 Energy Statistics Data Browser (2014). See www.iea.org/data-and-statistics?country=WORLD&fuel=Energy%20supply&indicator=Total%20primary%20energy%20supply%20(TPES)%20by%20source
15 Fernandes, "Ecosocialism from the margins" (2020a), 142.

References

Bonente, B. I. (2016). *Desenvolvimento em Marx e na teoria econômica: por uma crítica negativa do desenvolvimento capitalista.* Niterói: Eduff.

Esping-Andersen. (1990). *The three worlds of welfare capitalism.* Princeton: Princeton University Press.

Fernandes, S. (2020a). Ecosocialism from the margins. *NACLA Report on the Americas, 52(2).*

Fernandes, S. (2020b). Una visión ecosocialista para la descarbonización en América Latina. *Tricontinental, 52(2).*

Löwy, M. (2014). *O que é o ecossocialismo?* São Paulo: Cortez.

Marques, L. (2020). *Capitalism and environmental collapse.* New York: Springer Publishing.

Marx, K. (1982). *Capital* (Vol. 1). London: Penguin Books.

Marx, K., & Engels, F. (1976). *Collected works* (Vol. 5). New York: International Publishers.

Marx, K., & Engels, F. (1980). *Collected works* (Vol. 11). New York: International Publishers.

15 The case for a Leninist ecology

Considering everything presented throughout the book, it is safe to say that Marxism has much to offer to the debate of the ecological crisis and its many dimensions. It has a vast and rich literature made of very fruitful reflections both for the critique of capitalism and for the prefiguration of a post-capitalist society in an effective coevolutionary relationship with nature.

This should come as no surprise. It would be strange if a tradition of critique of capitalism, such as Marxism, did not generate a profusion of ecological reflections precisely when the destructive dynamics of this society are at their highest. What might come as a legitimate surprise is the more or less generalised deficit of action. The ecological critique of capitalism undertaken by countless Marxist thinkers demonstrates in depth not only the ecological unfeasibility of this society but also the urgency of overcoming it. However, this clear scientific awareness has not yet been able to catalyse a *praxis* with the appropriate sense of urgency.

In his 2021 book, Andreas Malm dedicates many pages to finding an explanation for this mismatch between the content of the already consolidated critique and the level and intensity of mobilisation.[1] Although he does not claim to offer a definitive explanation, the author seems to conclude that action falls short of discourse because of an undue insistence on direct action methods based on non-violence. According to him, the prevailing assessment that more incisive methods would alienate necessary support from society prevents, from the start, the required escalation of action aimed at translating awareness of urgency into effective urgent practice.

At first, this diagnosis seems to be overly based on recent experiences of civil disobedience movements, especially those in Western Europe. In any case, even momentarily overlooking the possibly incomplete nature of this understanding, it is easy to see that it points to a deficit of *action*. This chapter draws on the same author's previous book to explore another aspect of this current condition, suggesting the parallel existence of an *organisational* deficit.[2]

DOI: 10.4324/9781003512516-18

Considering recent Brazilian history, there have been several moments when the left came together around prominent agendas. *There will be no World Cup! There will be no coup! There will be no expenditure ceiling! They'll only arrest Lula over our dead bodies! There will be no pension reform! With the end of emergency aid, people will rise! The vaccine revolt is coming! The oxygen revolt is coming!*

The most striking common feature of all these statements is not that they have all been defeated. The most striking common feature is the expectation: "This is not possible! Now people will react!" In other words, the left very often manifests itself publicly as if counting on a spontaneous explosion of outrage and revolt. Indeed, there is always some level of organisation behind these slogans. However, two aspects of the necessary level of organisation seem to be absent.

The first is that the formation of a cohesive force, capable of intervening at decisive moments so that its slogans are not stillborn, is the product of an intentional and diligent process. The energy to catalyse a revolt in the desired direction must be built up. It does not spring *spontaneously* from everyday struggles, however valuable and necessary they may be. In Gramsci's words:

> The decisive element in every situation is the permanently organised and extensively prepared force that can move forward when a situation is deemed favourable (and it is only favourable to the extent that this force exists and is endowed with combative ardour). Therefore, the essential task is to dedicate oneself systematically and patiently to building up this force, developing it, making it increasingly homogeneous, compact, and self-aware.[3]

The second aspect that is missing is that the desired revolts also need to be built up, even before the opportune moment for their outbreak presents itself. Denunciation on social networks, in the streets, and through institutional channels plays a part. But this takes place at the retail level. The means for wholesale denunciation and agitation on an industrial scale must be built. In this regard, there are countless moments in Lenin's thinking, but the "fierce struggle against spontaneity" that he undertakes in *What is to be done?* deserves special attention. After arguing at length that spontaneity often places the working class "under the wing of the bourgeoisie," Lenin says the following:

> The upsurge of the masses proceeded and spread with uninterrupted continuity; it not only continued in the places where it began, but spread to new localities and to new strata of the population [...]. Revolutionaries, however, lagged behind this upsurge, both in their "theories" and in their activity; they failed to establish a constant and continuous organisation capable of leading the whole movement.[4]

Frequently, a debased spontaneity is defended on the grounds that the given conditions limit action. But in the passage quoted, Lenin rightly points out that this kind of position "misses the boat," even when insurrectionary energy spreads throughout society; even when a vaccine revolt or an oxygen revolt, for example, seems the most prominent and inevitable outcomes.

And if there is one arena in which the left cannot afford to "miss the boat," it is the climate arena. If there is one struggle in which the left cannot give in to the slow and bumpy pace and direction of spontaneity, it is this one. Since the given conditions are hardly (if at all) conducive to the kind of uprising that would be necessary, on the scale and in the direction required, then two urgent tasks lie ahead: to build the conducive conditions on the one hand and to be adequately prepared when they (suddenly) present themselves, on the other.

Being prepared, in this case, means not only systematically and patiently building the capacity for effective intervention but also, in many ways, prospecting the future. It means understanding that some very dramatic impacts of climate change that are already underway will begin to hit human societies with a ferocity and frequency for which there is still no satisfactory answer. It means comprehending that this uncontrollable disruption of "normality," which happened in a homoeopathic dose during the Covid-19 pandemic, will tremendously increase political volatility (sometimes episodically and possibly causing permanent level changes). It means realising that the pillars of capitalism (but also those of society in general) will be shaken violently.

In his 2018 book, *The Progress of This Storm*, Andreas Malm shows us that, unlike the left, the field of order has been diligently preparing.[5] The individual initiatives of multi-millionaires trying to create their own Noah's Arks are not even the most worrying part. Malm presents, for example, a US intelligence report on global threats which states: "Extreme weather events (floods, droughts, heat waves) will increasingly disrupt food and energy markets, exacerbating State weakness, forcing human migrations, and triggering riots, civil disobedience, and vandalism."[6] In the 2019 report, the list of imminent risks increases considerably: in addition to floods, droughts, and heat waves, fires, rising sea levels, soil degradation, pandemics, ocean acidification, mass migration, and geopolitical tensions are mentioned.[7] None of this is presented as a set of predictions for the distant future. These are ongoing processes that are already threatening (according to the same report) infrastructure, health, and water and food security. This is just one emblematic example, but it is clear that the left, in the broadest sense, is disproportionately ill-prepared.

Reflecting on this inertia, Malm proposes that Marxist ecological thinking has advanced a long way towards understanding the inescapable ecological unfeasibility of capitalism and that now a kind of "ecological Leninism" is needed.[8] This ecological Leninism would be structured around at least three basic principles.

The first is converting a crisis of symptoms into a crisis of causes. In other words, we need to understand Marxist ecology *in practice* and understand that as long as the logic of capital rules our lives, all possible ways of responding effectively to the challenges posed by the climate emergency will be blocked. We need to defeat the causes.

In line with this insight, Chapter 12 on geological voluntarism attempts to show how struggles oriented only towards making small advances or simply buying time are chronically incapable of achieving even the limited results they aim for. The destructive drive of capital is so great and so structural that small advances on the most superficial symptoms of this destruction are always neutralised and run over by the expansive impetus of the system as a whole.

The second basic principle to structure the ecological Leninism proposed by Malm is that there cannot be any kind of procrastination. It is essential to speed up our mobilisation. If little indicates that this speed can be achieved, it is essential to create the means to achieve it. The era of gradualism is over, says Malm, and everything discussed since Chapter 11 on climate change, illustrates the level of urgency involved very concretely.

The third and final principle is to seize any opportunity to take the reins of the State and force it into another direction: from a position of collaboration with the march of catastrophe to the kind of meticulous planning that will be required in the very near future. Certainly, Malm is not referring (any more than Lenin would) to possible electoral victories and the rise of reformist left-wing agendas in the bourgeois State. Both authors are fully aware that even well-intentioned governments face constraints that prevent them from acting at the root of problems. Taking the reins of the State has a revolutionary meaning here.

Turning once again to Gramsci, it is true that, when taking the wheel of the bourgeois State, a left-wing government, supported by popular pressure, may eventually obtain concessions from the ruling classes that mean specific and *momentary*[9] progress on the ecological agenda. However, Gramsci himself would warn:

> there is no doubt that such sacrifices and compromise [by the ruling classes] *cannot involve the essential*, since if hegemony is ethical-political, it cannot fail to be economic as well, it cannot fail to have its foundation in the decisive role that the ruling group plays in the decisive core of economic activity.[10]

In other words, the mere possibility of managing the bourgeois State does not bring along the possibility of "involving the essentials." Therefore, as long as this is the horizon when controlling the State, the following dilemma is inescapable: on the one hand, for the ecological agenda to be politically viable, it would have to be ineffective in terms of what is essential; on the other hand, if this agenda pretends to be effective, carrying out all the necessary transformations, it becomes politically unviable, unless it takes on a consciously disruptive, revolutionary character.

Even though these are just general principles, they convey ideas that must guide our organising efforts as we advance. Humanity has never faced a crisis like this. Dispersed struggles, with a horizon limited to everyday emergencies, are chronically insufficient. The climate crisis will not go away as the most recent crisis did. It will only get worse. The more the left places its bets on the spontaneous explosion of the masses, the further it is from safeguarding a habitable planet. It is impossible not to insist on Malm's question: "At what point do we escalate?"[11]

Throughout this book, Marxist ecology has been presented as a body of theory which is far from monolithic, offering a wealth of possibilities for a rigorous ecological critique of capitalism. As the natural processes that shape the ecological crisis worsen and, with this, impose increasingly imminent and severe risks to the conditions of life on the planet, a growing number of Marxists are pointing out that, alongside the critical power of our theoretical reflections, there is a more or less pronounced deficit of action.

On the one hand, a clear recognition of the severity of the situation has yet to spread sufficiently among Marxists who do not focus their work and research on environmental issues. This is the easiest gap to fill. On the other hand, as Malm points out, there seems to be an insistence on tactics that rule out more incisive measures that could be part of the list of alternatives in an eventual escalation of interventions that are consistent with the understanding of the ecological unfeasibility of capitalism and the imminent collapse of some natural systems that would threaten not only this civilisation (which must be overcome) but also the very existence of our species.

Without disagreeing with this diagnosis, I propose that the insufficient circulation of ecological critique among Marxists and this deficit of action pointed out by Malm intertwine with a deficit of organisation. In this sense, Malm's insight into using Lenin's revolutionary thought and experience to identify ways of overcoming these gaps could prove particularly fruitful. The next chapter takes a step in this direction.

Notes

1 Malm, *How to blow up a pipeline* (2021).
2 Malm, *Corona, climate, chronic emergency* (2020).
3 Gramsci, *Cadernos do cárcere* [*Prison notebooks*], vol. 3 (2014), 46.
4 Lenin, *Collected works*, vol. 5 (1977), 396–397.
5 Malm, *The progress of this storm* (2018).
6 US Intelligence Community, *Worldwide threat* (2013), 9.
7 US Intelligence Community, *Worldwide threat* (2019), 21.
8 Malm, *Corona, climate, chronic emergency* (2020).

9 To understand why this is necessarily momentary, it is enough to remember what was said in the presentation of the first principle of ecological Leninism. For a more detailed demonstration, See Sá Barreto, "Mudanças climáticas e a tarefa dos ecossocialistas: pelo abandono do voluntarismo geológico" ["Climate change and the task of ecosocialists: abandoning geological voluntarism"] (2021).
10 Gramsci, *Cadernos do cárcere* [*Prison notebooks*], vol. 3 (2014), 49.
11 Malm, *How to blow up a pipeline* (2021, Kindle edition).

References

Gramsci, A. (2014). *Cadernos do cárcere* (Vol. 3). Rio de Janeiro: Civilização Brasileira.
Lenin, V. I. (1977). *Collected works* (Vol. 5). Moscow: Progress Publishers.
Malm, A. (2018). *The progress of this storm.* London: Verso.
Malm, A. (2020). *Corona, climate, chronic emergency: war communism in the twenty-first century.* London: Verso.
Malm, A. (2021). *How to blow up a pipeline.* London: Verso.
Sá Barreto, E. (2021). Mudanças climáticas e a tarefa dos ecossocialistas: pelo abandono do voluntarismo geológico. *Economia e Sociedade, 30*(1), 211–234.
US Intelligence Community. (2013). *Worldwide threat assessment of the US intelligence community.* Retrieved from www.chds.us/ed/category/all/worldwide-threat-assessments/
US Intelligence Community. (2019). *Worldwide threat assessment of the US intelligence community.* Retrieved from www.chds.us/ed/category/all/worldwide-threat-assessments/

16 Revolutionary politics in face of the ecological crisis

If this diagnosis maintains that capitalist society is ecologically unviable and can only deepen the destruction and destabilisation of the planet's natural systems, then any policy proposed, *even at a tactical level*, needs to be consistent with this understanding.

At other times the force of the working class has indeed proved capable of obtaining substantial concessions from capital, even if these concessions are tremendously concentrated in the centre of the system and, even there, temporary. However, on the ecological front, the tasks are so monumental, so urgent, and so decisive that any small step in the right direction takes us into territory where capital cannot concede. Regardless of how the virtues and limits of past reforms are assessed, simply transposing the possibilities and impossibilities revealed there into the ecological sphere is a mistake.

The reflection on the ecological crisis must be specific and capable of laying the theoretical foundations for a revolutionary movement that *cannot afford not to be victorious*. It is easy to see that when the edifice of revolutionary practice stands on shaky ground or fragile pillars, any strategy devised will suffer from debilitating problems of origin.

Before anything else, therefore, it is essential to explicitly reject conceptions of tactics and strategy that lead us down dead ends. There are basically two ways of relating tactics and strategy that need to be criticised: a *purely temporal approach*, taking tactics as the short-term horizon and strategy as the long-term horizon, and another, a *possibilities approach*, taking tactics as the set of actions aimed at the possibilities of immediate success (or the urgencies of everyday life) and strategy as the set of actions aimed at an indeterminate future, still beyond the reach of possibilities.

A first problem, shared by both approaches, is that tactics and strategy appear as spheres of action related to dissociated sets of objectives. The urgencies of everyday life, on the one hand, and a dream of the future, on the other. The small victories that are insufficient (but possible) today, on the one hand, and the great victories that are necessary (but impossible at the moment), on the other.

DOI: 10.4324/9781003512516-19

It might not seem like such a problem at first glance, given how often tactics and strategy are discussed in these terms. However, it is crucial to recover the complete understanding that there is (and should be) *a subordinate relationship* between tactics and strategy. More critical than recovering this understanding, which has never wholly ceased to circulate, is to become capable of drawing *a political line that corresponds to it.*

Indeed, strategy should be conceived with the long term in mind. But not in the sense of aiming for a distant long term, disconnected from the present, and then turning all our energies to the hassles of everyday life. On the contrary, this extended time frame reaches into the future (which may be more or less distant) *but begins today*, in the present. Tactics must, therefore, be subordinate to strategy. The question that structures tactics *cannot be* "What can I (or do I have to) do today?". The question that structures tactics *has to be* "What can I (or do I have to) do today *in the service of strategy?*". Tasks are determined *by their relationship to the tactic rather than* by their location in time or the most immediate agendas. In the same way, tactics are determined by their relationship to strategy.

Two other problems may arise from a poor understanding of tactics and strategy. One specific to the "short/long term" version. Another specific to the "possible/not yet possible" version.

When confining tactics to the short term and strategy to the long term, arbitrarily dissociating them, one *inadvertently* makes the unusual mistake of conceiving the long term as a linear and necessary result of the short term. Something that could be said more or less like this: if we throw all our energy into the struggles of the present, the conditions for victory in the long term will be more or less automatically and spontaneously generated. So, strictly speaking, we would be building the victories of the long term by *simply embracing* all the struggles of the present that seem urgent to us *in the present.*

There are two critical mistakes in this understanding. First, the present will always be full of struggles and urgencies. Suppose the systematic, diligent, and persistent construction of the revolution is postponed when urgent struggles must be fought. In that case, it is clear and certain that this construction will be postponed *forever*. Second, among these urgent struggles of the present, there will be those that deviate from the revolutionary path, those that momentarily suspend the journey in this direction, and those that bring possibilities of effectively accumulating forces to move in this direction. If tactics are limited to the short term and dissociated from strategy, the conclusion *a priori* is that all urgent struggles must be waged. But if tactics are consciously subordinated to strategy, even urgent struggles are different in nature and, therefore, require qualitatively different reflection and engagement.

Finally, by confining tactics to what is *immediately* possible, one puts small victories within reach. Since, in this case, tactics are not explicitly and consciously subordinated to strategy, it is not asked whether or not these small victories can be accumulated until the eventual unleashing of a revolutionary process, whether or not these small victories, by their very nature, are fragile and subject to revocation, or even whether or not these small victories are illusory.

This is a crucial question. It is not advisable to assume that, at some point, it will be possible to bring together, all at once, all the conditions necessary for a break with capitalist society. When limiting tactics to the struggles that have the conditions for immediate success, instead of building the organisational capacities to be in a position to catalyse and direct a revolutionary process, small, compartmentalised victories are achieved. The hope is that engaging in these struggles will somehow ferment a revolutionary political consciousness. It is only fair to remember that there is some (albeit precarious) textual support for this expectation, even in Marx. However, if the "lesson" conveyed by these struggles is that capitalism can be better than it is, then the most likely outcome is that the consciousness generated there will have a counter-revolutionary character.

This debility is particularly evident in the discussions on transition, approached below.

The idea of transition is very present in ecosocialist thought and indicates the necessarily procedural nature of overcoming capitalist society. Detailing this process a little further, it is possible to say that we live in a world ruled by capital and advocate another possible world in which capital has been overcome; this new world, however, is not created out of nothing but is born from the guts of the current world; therefore, our *praxis* is faced with (and directed at) the parameters, structures, needs, and urgencies of the current world; hence the need for a transition program, i.e., an action program that, while struggling with (and in) the present, points to a possible future.

This is where the problems begin. The first and most obvious is to confuse *a transition program* with *the transition itself*. The analysis of one of the best-known transition programs reveals that the proposed set of struggles aims to make several advances from the point of view of the working class.[1] However, it does not claim that these advances alone will bring about a transition from the current society to the future society. In this sense, the transition program *does not operate a transition*, waiting for a revolutionary period. Much less does it operate a transition that would make it possible to dispense with the revolution for as long as possible. If carried out successfully, the transition program fulfils the task of straining the present to its breaking point. In other words, it fulfils the task of unleashing the revolution, *not waiting for it*. It is the revolution, in turn, that really opens up the possibilities for an effective transition.

In conclusion, an ecosocialist transition program *cannot foster illusions that the neuralgic ecological problems of capitalism can be solved within capitalism*, i.e., without challenging and destroying its fundamental parameters of reproduction to the marrow. An ecosocialist transition program must outline a path for building *a revolution that leads to an ecosocialist direction.*

At first, situating *transition* and *revolution* in time may seem like pure theoretical pedantry. However, concerning the issues that interest ecosocialist thought, a proper understanding of what is possible *before* a revolution and what becomes possible only *in the midst* of a revolution is decisive.

There is no need to exclude *a priori* the possibility that, for a series of critical social transformations, transitions occur within the current order. However, the issue of the ecological crisis calls for a specific analysis, as emphasised. At the heart of the debate on the ecosocialist transition is the urgent need to carry out an energy transition capable of decarbonising our productive activities. As covered in Chapter 14, this brings a series of monumental challenges, all irreconcilable with the reproduction of capitalist society. This point may be once again addressed using a method of demonstration by absurdity.

Given that the world today is profoundly unequal, an energy transition that equally imposes the abolition of fossil fuels on all nations would undoubtedly be unjust, leaving huge fractions of humanity trapped in a situation of material hardship even more deplorable than today. That is why ecosocialists advocate a just transition, which implies the need for the most powerful nations in the system to make the most significant sacrifices. Is it plausible that the most economically and militarily powerful nations on the planet, where all the largest and most important gravitational centres of capital are located, would be willing to take the lead in this process? Would they passively watch it unfold, driven by reformist political will and popular pressure?

Another relevant point is that our current capacity to supply energy from so-called renewable sources is comparable to the primary volume of energy demand in the early 1950s. Therefore, the rapid elimination of fossil fuels would require accommodating the demand for energy to something compatible with that level of supply. Naturally, this would require a colossal shrinkage not only of residential consumption but also of energy consumption in various productive activities. It would also imply an extraordinary downturn in productive/economic activity. Is it plausible to imagine that the global capitalist economy would tolerate an abrupt, immense, and permanent shrinkage in economic activity?

A similar question may be asked about consumption. Right away, it is easy to see the need to eliminate the consumption of useless knick-knacks of various kinds. It is easy to see that even the consumption of some essential items may eventually need to be rationed or suspended. It is easy to see that items produced for consumption must last longer, not shorter. Faced with such trivial observations, one may ask: is it plausible to imagine that a production system that increasingly, rapidly, and uncontrollably dumps mountains of products on the market would tolerate any moderation in global consumption?

To assume, *in these cases*, the idea of transition as a process that *precedes* a revolutionary process, it is necessary to presume that these tasks are all *still achievable within the framework of capitalist society.* Thus, it is supposed that the transition would be a goal that, once achieved, would at least guarantee the ecological conditions of life on the planet so that, finally, at some point in the future, the revolution may be unleashed. Admitting this implies admitting, even unwillingly, that capitalist society can resolve and overcome its destructive and destabilising character in the energy/climate sphere. Unfortunately, many brilliant ecosocialist comrades fall into this trap to a greater or lesser extent.

Therefore, the apparently simple task of conceiving the transition as either a process that *precedes* the revolution or a process *made possible* by the revolution has very profound implications on how to approach tactical questions. On the one hand, if it is assumed that necessary transitions can still be achieved under capitalism, the transition program will be full of reforms. Reforms that may even, in the end, bring timid progress. Considering everything covered in this book, however, it is possible to affirm that the magnitude of the necessary advances would eclipse them. On the other hand, if it is assumed that the transition can only be disputed in the midst of a revolutionary process, the transition program will be populated with struggles that, although formulated in terms of the urgencies of the present, will have the conscious intention of tearing the fabric of capitalist society to its breaking point. They will aim to catalyse the revolution. And talking about an ecological and climate crisis with severe implications for life on the planet, they will have the utmost urgency to do so.

Understood in this way, the question of transition and the transition program requires us to look carefully, without ready-made formulas, at the relationship between minimum and maximum programs.

From the outset, it is easy to see that the concept of minimum encompasses many different ideas with different tactical implications, some of which are irreconcilable. It can refer to the minimum *that can be* achieved, given the present conditions. A more complete version can refer to the minimum that can be achieved in the present moving *towards* the maximum program. In a worse version, it can refer to the minimum that can be achieved without breaking with the current system or causing any social instability. It can also refer to the *minimum necessary*, given the present urgencies or some extended time frame. Finally, it can refer to the minimum necessary, *given the maximum program*.

Amid this variety of minimums, two basic classes appear: the minimum *possible* and the minimum *necessary*. By not paying enough attention to these nuances, one may conflate possible and necessary without even realising it. This gives rise to the uncritical admission that what is necessary is possible, simply because it is necessary. And that what is possible is necessary, simply because it is possible.

Indeed, there can be a high number of tasks in which the immediately possible and the necessary coincide or have substantive intersections. This, unfortunately, is not the case with the tasks posed by the challenges related to the ecological and climate crises. Decarbonising our way of life is no trivial task. The quantitative and qualitative moderation of humanity's total consumption is no trivial task. Transitions that impose more sacrifices on rich nations than on poor nations are no trivial task. The accumulation of sufficient force to impose transitions on the reactionary and conservative field is no trivial task.

The difficulties are not limited to the undeniable colossal magnitude of these tasks. The most important difficulties derive from the *impossibility* of accomplishing them outside of a process of breaking with the order of capital. Even though this statement may seem controversial at first glance, this entire book has been, to some extent, an effort to demonstrate it rigorously. When discussing the minimum, two things stand out.

First, if the minimum necessary is impossible, then what is possible will fall short of what is necessary. And if the minimum necessary refers to the minimum to guarantee the survival of our species, then falling short is, quite literally, deadly. And given the speed of the ongoing process of climate and ecological destabilisation, the consequences of falling short come very quickly. Directing all of our energy to fight towards what is possible, simply because it is possible *now*, is not realism or pragmatism, it is not caring about the most urgent needs of the working class. Although it may seem that way, such uncritical dedication to what is *possible* now amounts to historical procrastination.

Second, it is not trivial to say that such an indispensable set of tasks is impossible. This usually provokes a strong (and legitimate) reaction against what sounds like a *fatalistic* statement with all its negative implications. Therefore, it must be emphasised tirelessly that this impossibility is, at least for the time being, confined to capitalist society. In other words, all the necessary actions to safeguard a habitable planet, even the *bare minimum*, are impossible under *capitalism*.

It is capital that cannot give up fossil fuels, not humanity. It is capital that cannot give up growth, not humanity. It is capital that cannot give up countless destructive productive activities, not humanity. It is capital that needs to mobilise all our technological prowess to ensure growth at any cost, not humanity. It is capital that needs the space of entire forests to occupy them with intensive monocultures, not humanity. It is capital that needs to ensure that consumption is ever higher, more varied, and faster, not humanity.

Humanity, on the other hand, needs to make the energy transition, reduce material demand on the planet, eliminate destructive sectors, mobilise technology and technological progress to moderate our ecological impact, guarantee the full existence and reproduction of biomes and ecosystems, and moderate global consumption.

Having said that, it is patently clear that the necessary actions are irreconcilable with the primary needs of capital. This inability of capitalist society to accommodate the minimum program (understood here as the minimum necessary) in any way forces us to think of *a minimum before the minimum.*

It could be formulated in the following terms. The bare minimum involves making a series of transitions (energy, agriculture, industry, cities, etc.), and the most urgent and within reach is energy. However, even this transition is unrealisable through a set of reforms. It is, therefore, impossible under the rule of capital. If it is really necessary (if it is, in fact, a matter of life and death), it must be made possible. Since it is the reproductive dynamic of capital that makes it impossible, only the overcoming of capital brings it within the field of possibilities. In other words, only through a revolutionary process can *these impossibilities become possibilities*. If this reasoning is correct, it can be said, without exaggeration, without thoughtless impatience in the face of situations x, y, or z, that the revolution is our minimum program.

From a tactical point of view, there are crucial ramifications of this conclusion. As institutional politics within the bourgeois State has clear limits and shortcomings, the left should not bet all (or even most) of its chips on this path. This is a sphere where the trenches established will, at best, impose some friction upon the destructive appetite of capital.

If only a revolution can unlock the possibilities of truly effective action to tackle the climate and ecological crises, building the revolution must be the focus of efforts, always making the most of the conditions that exist today (even if they are scarce). Finally, understanding that some disruptive impacts are already inevitable, it is also crucial to prepare for them autonomously, outside the State.

In short, we must fight for the world we want, but knowing that we will have to find ways of dealing with a planet with an increasingly hostile nature, the result of centuries of destruction and global ecological destabilisation driven by the blind logic of capital. And we need, indeed, to modulate our patience, but based on the real urgencies we face, not based on an abstract ideal of patience. And we have to do this without the slightest attachment to the structures of the bourgeois society we live in.

Note

1 Trotsky, *Death agony of capitalism and the tasks of the fourth international* (1981).

Reference

Trotsky, L. (1981). *The death agony of capitalism and the tasks of the fourth international: the mobilization of the masses around transitional demands to prepare the conquest of power: the transitional program.* New York: Labor Publications.

17 Epilogue

Senile capitalism in pandemic crisis

After going through the ecological critique of capitalism, the defining characteristics of ecosocialist thinking, and some debates on tactics and strategy, this final chapter turns to a complementary double exercise. First, to point out that, beneath the more or less frequent succession of economic crises, there is a much more persistent and less oscillating trend to deepen material demand on the planet and, therefore, ecological destruction. Second, to point out that, in addition to the possible parallels between 20th-century capitalism and contemporary capitalism, and in addition to the characteristics of this system that have been reproduced and preserved throughout this time, it is also possible to highlight the emergence of a series of fundamentally new elements (either in scale or quality), which can be mapped by investigating the links between the structural crisis of capitalism, the ongoing ecological crisis, and the health crisis triggered by the Covid-19 pandemic.

From a general/global point of view, the destructive character of the dynamics of capital has already been pointed out and demonstrated exhaustively in Marxist literature. Even the ecological dimension of this destructiveness has been receiving reasonable attention for at least a couple of decades. The Covid-19 pandemic, declared by the WHO at the beginning of 2020, added another focus of attention to this ecological critique of capitalist society.

Authors such as Malm, Wallace, Foster, and Suwandi have already presented, each in their own way, solid arguments connecting the most immediate drivers of the health crisis to the unstoppable capitalist drive for expansion.[1] Before looking into the most decisive points of this relationship, it is important to be more precise about this drive for expansion.

It is no exaggeration to say that the main accounting concept that informs common sense about the growth trajectory of economic activity is GDP and its periodic variation. It also seems safe to admit that even in common sense there are conceptions about the cyclical nature of the economy, which successively goes through phases of heated production and consumption, slowdown, and eventual retraction. At the same time, the precarious recent history of economic

DOI: 10.4324/9781003512516-20

growth in several central countries (and even in peripheral countries) has already prompted reflections on a supposed secular stagnation.[2]

Taking only these elements into account, the destructive nature of the capitalist drive for expansion remains out of sight. In the case of GDP, this metric focuses on the social form of wealth, on value. The material dimension of wealth (and, therefore, of the productive activities that generate it) is overlooked. Regarding the conceptions of the undulatory movement of economic activity, it is easy to lose sight of the fact that the succession of cycles can have directionality. And even bearing this in mind, the actual performance of economies could lead one to believe that this directionality is, in all likelihood, only timidly increasing. All things considered, does it make sense to talk about a persistent destructive drive for expansion?

Marx himself carefully demonstrated, especially in Chapter 23 of Book I of *Capital*, that accumulation goes through moments of acceleration and deceleration.[3] Furthermore, his discussion of the overproduction of capital reveals that accumulation goes even through moments of retraction. However, it is crucial to point out that this whole movement describes the trajectory of wealth as value. As mentioned before, in capitalist society, wealth is value and use value. Therefore, it is important to know what kind of movement material wealth undergoes amid the oscillations of wealth in value form.

The shortest way to do this is by proposing a brief exercise on the impact the rise in productive forces has on material wealth and wealth in value form. The first thing to consider is that the increase in productive forces implies a decrease in the volume of work that is socially necessary for production. As a result, the value of each unit produced decreases. From the point of view of value, thinking only of this unit, there has been a decrease. From a material point of view, nothing has changed. The unit continues to be produced, and the production process continues to process the same volumes of raw and auxiliary materials, generating the same volumes of waste. Only production time is saved. It is already clear that the pattern of movement of material wealth does not follow the pattern of movement of wealth as value. From a material point of view, the same useful object continues to be produced, the material demand for productive resources remains the same, and so does the potentially polluting impact of productive activity. From the point of view of value, taking only the unit produced as the object of analysis, there has been a retreat.

Shifting the focus from the unit produced to the *total* number of units produced, could this total remain the same, in which case the material impact of production would at least remain stationary? No. As the value of the unit produced has decreased, the production of the same number of units as before would imply a lower total value than before. Thus, capital guarantees the conditions for its expanding reproduction only by increasing the number of units produced enough to compensate for the decrease in the value of each unit, allowing for an increase in the total value produced.

To make this exercise more detailed, let's analyse three possible scenarios: (i) the level of production increases to such an extent that it guarantees an increase in the sum of value produced; (ii) the level of production increases to such an extent that only the same sum as before is guaranteed; (iii) the level of production increases, but to an extent insufficient to compensate for the decrease in unit value so that the total value decreases. Each of these scenarios, respectively, represents expansion, stationarity, and decrease of wealth as value. What happens from a material point of view? In all cases, production increases. In all cases, therefore, the volume of useful objects produced increases. The magnitude of the material demand on productive resources increases. The volume of waste produced increases. In other words, from a material point of view, *the movement carried out by capital-led production is persistently increasing*, unlike the pattern of movement in the value dimension.

The growing material demand on resources, biomes, and ecosystems pointed out by Malm, Wallace, Foster, and Suwandi (and many others) is, therefore, much more than just an empirical observation or an understanding of the *history* of capitalism. The application of the categories of the critique of political economy leads to the conclusion that the society of capital can present no other pattern of pressure/demand on the planet than this intensive and extensive growth.

The dynamics briefly reviewed structure the relationship between humanity and nature in such a way that the socio-metabolism corresponding to capital exhausts, destabilises, overloads, and destroys nature. The ecological crisis is thus an inseparable material expression of the development of this society. But how do this predatory relationship and this multifaceted crisis connect to the health crisis?

At first approach, at least three trends appear as catalysts for epidemic episodes: the advance of the agricultural frontier, the concentration of livestock populations, and large urban concentrations.

As Wallace and Malm show us, wild environments are true reservoirs of viruses in organic equilibrium with their hosts. As the devastation of the soil caused by capitalist farming requires the opening up of new areas for cultivation or grazing, the area of contact between human populations and wild animal populations expands and becomes more porous. As a result, the risk of zoonoses increases in direct proportion to the frequency of this contact. Once zoonoses have occurred, the virus may cause the disease in the new host, be it a human being, a domestic animal, or livestock.

This brings us to the second point: even in the (increasingly rare) cases where the activity is extensive, the genetic homogenisation that interests the economic exploitation of these animals leaves these populations more vulnerable to the rapid spread of diseases when a virus eventually jumps from a wild host to an animal from the herd. Once this happens, the risk of humans becoming infected increases since the disease is no longer at the border between a wild environment and a human settlement. Now, it is in a production environment in which, regardless of the level of productivity, workers circulate. Depending on the specific circumstances, these workers may have to travel short distances between their workplace and a small urban centre every day. They may also have to travel long distances in large urban centres and on crowded public transport. One way or another, when a disease spreads rapidly through a herd, workers in contact with the sick population become potential vectors of transmission to human populations clustered in cities.

Finally, we come to the third point. Depending on the mechanism of contagion, physical proximity between infected and uninfected individuals is a key variable. At least three characteristics of large urban centres make them powerful "routers" for spreading epidemics: agglomerations, large distances that need to be travelled routinely, and speed of movement.

These are features of contemporary capitalism that increase the risk of viral outbreaks and transition from outbreaks to epidemics. Insofar as the expansion of the agricultural frontier is not a mere deviation from the course, the way it integrates the processes that generate zoonoses, outbreaks, and epidemics can (and should) be traced back to this system's most essential operational logic. Insofar as the genetic homogenisation of herds is the direct result of cost-control and revenue-boosting practices, the role it plays in the outbreak of epidemic events should be traced back to this system's most essential operational logic. Insofar as the separation between town and country is the "natural" landscape of the world shaped in the image and likeness of capital, the abundance of conditions conducive to epidemics in large urban centres should be traced back to this system's most essential operational logic.

The three points covered form a set of conditions that are conducive to the emergence and local spread of diseases in humans. However, the link between the structural, ecological, and health crises is not limited to them. Let's take this journey again, from the economic to the sanitary, passing through the ecological, to map not only favourable conditions but also important drivers of these crises.

Marxist thinking on economic crises includes several different perspectives, which are not always easy to reconcile. Amid this variety, however, two conceptions always seem to be present. On the one hand, there is the recognition that the conditions for the crisis are created during the period of vigorous accumulation and that the conditions for the recovery are created during the crisis. In other words, the recognition of a cyclical movement. On the other hand, the idea that the capitalist mode of production is tremendously plastic and has shown itself capable of circumventing barriers, absorbing resistance, and stifling contradictions in order to preserve itself as essentially the same.

Without disagreeing with these conceptions, two details are important. First, the cyclical movement is not sinusoidal. The succession of waves can characterise an increasing or decreasing direction. The amplitude of the waves can increase or decrease in a more or less unpredictable way, or it can increase or decrease progressively. Finally, the recognition of this cyclical nature and that one phase creates the conditions for the next is not enough to conclude that there is necessarily an eternal return to the boom phase. We cannot even conclude that the *possibility* of a return to the boom phase is preserved indefinitely.

Second, the plasticity of a society ruled by capital is not unlimited. Not every barrier is surmountable within the reproductive parameters of capital. Furthermore, as it changes to remain the same, capitalism leaves behind configurations that have become unviable, i.e., configurations to which it cannot return. It is impossible not to think of Meryl Streep and Goldie Hawn's characters in the movie *Death Becomes Her*. Although immortal, they deteriorate as they have to make interventions to preserve themselves. The exercise of plasticity explores new possible configurations but exhausts the previous ones. In other words, as it changes to remain the same, the reproductive alternatives of capital become increasingly narrow. The exercise of plasticity stiffens it.

These two additions reveal that the cyclical nature of capitalist society is perfectly compatible with an undulatory movement of increasing frequency and/or with ever-smaller peaks and ever-deeper valleys. Furthermore, this plastic character of capital is perfectly compatible with a stage of development in which its fundamental contradictions reach a climax. This is precisely the framework of what can be called a structural crisis.

The issue of structural crisis is very controversial and often confused with the issue of a supposed imminent collapse of capitalist society, or a supposed obsolescence of the law of value, or a supposed obsolescence of the Marxist theoretical legacy for understanding this reality. None of these things is the point here. The point is the fundamental contradiction of this system, which drives the absolute development of the productive forces (making living labour increasingly superfluous) while maintaining labour as the foundation of the social form of wealth, i.e., value.

In the historical development of capitalist society, this contradiction is "resolved" in a way that suits the accumulation of capital. However, this "resolution" does not extinguish the contradiction but shifts it to an even higher level of tension. In concrete terms, the absolute development of the productive forces, driven by the logic of relative surplus value, makes human labour relatively superfluous. At first, this could be an obstacle to accumulation, given that labour is the substance of value. This obstacle, Marx shows, is spontaneously circumvented because the growth in the scale and scope of capitalist operations guarantees an expansion in its capacity to absorb more labour, even amid a proportional decline in the participation of living labour in the production process.

The classical theories of imperialism provide the insight that this mechanism for resolving the contradiction manifests itself concretely as a spreading of capital throughout the world, both in a spatial/geographical sense and in a purely social sense, with the penetration of the logic of capital into every crevice of everyday life.[4]

However, Marx himself already anticipates two sensitive issues: on the one hand, this blind dynamic of rising productive forces in itself guarantees that the new branches of production that emerge in the course of the spread of capital tend to enter the scene with a diminished capacity to absorb living labour. This new capital employs proportionally less labour. On the other hand, even the productive apparatus already in operation comes to the end of its useful life. When it is replaced by a new, more productive one, previously employed workers become unemployed.[5]

Marx presents this reasoning with a much greater emphasis on the impact of these movements on the working class. But it is easy to see that these same movements describe a kind of compensation mechanism that tends towards exhaustion. The more capital does without living labour in the production process, the greater the scale and scope of accumulation must be to guarantee the conditions for increasing extraction of surplus labour.

When it comes to imperialism, there are also tendencies towards exhaustion. From a spatial/geographical point of view, the more capital spreads its tentacles across the planet, the fewer unexplored corners remain to be colonised. From a social point of view, the more the logic of mercantile production is installed in (and takes hold of) every moment of life in society, the fewer moments remain untouched.

At this point, it is crucial to realise that the gradual exhaustion of this compensation mechanism does not put a brake on the movement as a whole. Quite the opposite: the quest to compensate for the relatively growing absence of living labour by accelerating the suction of living labour occurs (and needs to occur) in an even more intense, comprehensive, and unbridled way.

An interesting parallel with a dying star can be drawn here. Very schematically, considering a star like our sun as an example, the thermonuclear fusion occurring inside the star will eventually completely exhaust its primary fuel, hydrogen. With the exhaustion of the outward pressure generated by the fusion of hydrogen into helium, the inward pressure will eventually raise the temperature to the level at which the fusion of helium into carbon takes place. As a result, the outward pressure generated by this new stage will cause the star to reach a size hundreds of times larger than its original size. Here is the right point for the analogy. Similar to the dying star, capitalist society reaches its widest and deepest extent (spatially and socially) the more chronically incapable it becomes of "burning" its primary fuel, human labour. In other words, the social metabolism corresponding to capital seems more totalising and unavoidable precisely in its most senile stage.

In concrete terms, this dynamic manifests itself in various ways. Two of them are emblematic for understanding the connection between the structural crisis, the propensity to health crises, and the ecological crisis: the dense network of very long supply chains that spreads across the globe and the hypertrophy of the circuits for the valorisation of fictitious capital.

As stated by Foster and Suwandi:

> These commodity chains make up the main material circuits of capital globally that constitute the phenomenon of late imperialism identified with the rise of generalized monopoly-finance capital. In this system, exorbitant imperial rents from the control of global production are obtained not only from *global labor arbitrage*, through which multinational corporations with their headquarters in the center of the system overexploit industrial labor in the periphery, but also increasingly through *global land arbitrage*, in which agribusiness multinationals expropriate cheap land (and labor) in the Global South so as to produce exportable crops mainly for sale in the Global North.[6]

The key player here is the word *material*. In research and debate on globalisation, there is often an emphasis on the abstract dimension of both wealth and the movement of capital. Rarely do we pay due attention to the counterpart of this movement in terms of forest, water, mineral, and chemical resources, among others.

At this point, elements of nature are referred to as "resources" not only because this is common sense but because it indicates how capital organises a specific relationship between humanity and nature: a relationship of instrumental appropriation. In this way, as well as highlighting the pressures of exhaustion of crucial resources across the planet generated by globalisation in its material dimension, we also bring together elements for a more systemic approach to the same issue. To this end, Foster's contributions are essential.[7]

As presented in Chapter 2, the author's revival of the Marxian discussion about the rift in the metabolism between humanity and nature that characterises capitalist society has a precise meaning. It is about the disruption of the soil nutrient cycle caused by the spatial separation between town and country, leading to the destruction of soil fertility conditions, on the one hand, and the production of a constant and growing flow of pollution in cities (at first), on the other. In this chapter, however, the focus is on the process that takes this separation to global scales and the rift to its maximum. It is not just the fertility of the land that is being depleted. It is entire biomes, species, and ecosystems. It is not just large urban centres that are buried by pollution. It is rivers, seas, oceans, the country, the underground water reserves, and the atmosphere.

The imperialism that confronts us today, therefore, is not only accelerated by the increasingly demanding requirements for the reproduction of a dying capitalism. It also deepens and accelerates the destructive tendencies that generate and accelerate the ecological crisis.

In addition, it is clear that the material hyperconnection of production activities, which means that the production of a simple commodity travels to countless corners of the globe, acts as an extremely powerful transmission belt for microorganisms that cause pathologies or for the carriers of these pathologies themselves. In other words, it acts as a force that increases the likelihood of outbreaks or epidemics turning into pandemics.

Again, there are two manifestations to emphasise at this point in the text. The first, concerning globalisation, indicates that the natural, unavoidable, and undeniable appearance of capitalist society is a symptom of its senility. Also, this senility, instead of moderating the destructive appetite of capital, accelerates it. The second concerns the hypertrophy of the logic of valorisation of fictitious capital.

Strictly speaking, the argument presented so far regarding the structural crisis adds little to what Marx himself had already demonstrated in *Capital*. In Book III, the discussion of the tendency for the rate of profit to decrease and the overaccumulation of capital shows that, in the course of its development, capital is faced with increasingly demanding requirements to guarantee the conditions for its valorisation. What was added here is the deepening of these difficulties to the point where the crisis becomes (without excluding the possibility of oscillations) chronic.

The permanently decreasing capacity to guarantee the conditions for accumulation based on the surplus value produced forces capital to shift the dynamic axis of its accumulation towards the future. In other words, the logic of (expanded) reproduction of fictitious capital becomes increasingly fundamental. It is worth pointing out that this does not mean that there have never been other moments of hypertrophy of the financial sphere, which until today have always been succeeded by periods of return to a dynamic characterised by really active capital. What is specific to this moment is not being unprecedented but its irreversibility. In other words, the current protagonism of the financial logic only tends to deepen.

Thus, among the various peculiarities of *fictitious capital*, one is particularly interesting. Fictitious capital consists of a title of appropriation over social wealth that depends fundamentally on the expected flow of future surplus value, not on the value and surplus value *already produced*. From an ecological point of view, two consequences stand out.

First, this fictitious wealth will be more secure the closer the expectation of such a future flow is to being fulfilled over time. This inscribes an additional imperative determining the drive for growth in the general movement of the system. It is not just the objective of accumulation that is frustrated if growth is not enough to feed back into the logic of fictitious capital. A fraction of the current "wealth" backed by this type of capital evaporates when its material basis is not concretely confirmed.

This leads to the second consequence. Although fictitious capital refers to future surplus value (which may *or may not* be produced, depending on the circumstances), it commands purchasing power in the present. As such, the expansion of society's consumption capacity at any given moment is, at least in part, guaranteed by the expansion of fictitious capital. In a very literal sense, this allows humanity to anticipate productive and unproductive consumption. Putting this reasoning into a material key, it is as if we were consuming our future.

Taking advantage of the discussion on the irreversible role of the logic of fictitious capital, a third dimension of the articulation between the structural, health, and ecological crises deserves attention. The originating and accelerating determinants have already been discussed. The following paragraphs analyse how the trends behind the structural crisis mean that humanity is increasingly ill-equipped to deal with health crisis events and the ecological crisis.

As the conditions for the extraction of surplus labour become ever more rarefied in proportion to the magnitude of the capital that needs to be valorised, the general metabolism of the system needs to find ways of ensuring these conditions. This, on the one hand, leads to a flight into fictitious capital, as seen before. This flight creates, in addition to everything presented until here, monumental restrictions for humanity to carry out the transformations that would be necessary to suspend our disruptive ecological influence.

Chapter 14 explained that any ambition to decarbonise would imply, at the very least (thinking along extremely moderate political lines), not exploiting already known oil, gas, and coal reserves. However, although not yet materially exploited, these reserves *are already circulating, in the present, as fictitious capital*. Once again, a global veto on their exploitation would imply the sudden disappearance of trillions of dollars in financial assets. A system whose gears turn only if oiled by increasingly colossal volumes of fictitious capital will not tolerate something of this nature.[8]

But the very flight into fictitious capital reinforces, on the other hand, the need for a parallel "way out": to ensure at all costs that *all* the opportunities for extracting surplus labour that are still beyond the reach of capital *are brought within the reach of capital*.

The obsession with privatising public assets can be understood in this context. The obsession with deregulating the movements of fictitious capital in the global economy can be read in this context. Likewise, the tight straitjacket that the market places on the State and public policies. Finally, the voracity with which capital encroaches on public funds, labour rights, and the income of the working class.[9] One could also include in this same interpretation the obsession with just-in-time production mentioned by Foster and Suwandi.

Based on these insights, we can assess the health crisis of 2020 and 2021 from a less conjunctural or politicistic perspective. The outbreak of the Covid-19 epidemic in China (which turned into a pandemic within a few months) strongly demonstrated the need for idle capacity in hospitals and some industries. However, this kind of idleness is the enemy of capital, especially senile capital obsessed with *just-in-time* production. It also demonstrated the need to have the capacity to produce a range of equipment locally. However, the compulsion for cost efficiency demands that the bulk of the world's industrial production be concentrated in China. It has also demonstrated the need to suspend, even if temporarily, the movement of people, animals, and things (in cities and around the world). However, this kind of paralysis is the enemy of capital that is materially dependent on a dense network of very long supply chains. In fact, the whole debate that took place about whether or not to paralyse essential activities has turned out to be quite symptomatic.

Ordinary people modulated their positions according to what is or is not necessary to fulfil everyday human needs. They have lost sight of the fact that, in this society, the need that really matters, which subordinates all others, is that of capital valorisation. Given the network of absolute and global material dependence that today structures the reproduction of capital, everything is essential and, therefore, *nothing can stop.*

For the people who lost their incomes in the economic pandemonium that followed the health crisis, some kind of financial bailout would have been essential. But the bourgeois State's fiscalist and anti-inflationary fixation, objectively determined by the reproductive requirements of this dying capitalism, ensured that the bailout, even where it existed, fell short of what was needed.

The listing of these specific examples emphasises why humanity was impacted by Covid-19 in a general situation of profound unpreparedness, despite having the technical capacity available and reasonable knowledge of the risk of a pandemic for years. Not because we did not know what could and needed to be done but because what could and needed to be done was, to a large extent, incompatible with the reproductive requirements of capital.

It is precisely this kind of Gordian knot that is found in the context of ecological crisis. Everything we need to avoid the worst impacts of this crisis (and which is within reach from a technical point of view) is irreconcilable with the logic of capital. What is more, everything we need to do to adapt to the already inevitable and reasonably foreseeable impacts also displays this same characteristic: the already scarce and insufficient means to respond to the health and ecological crises are rapidly being eroded as the paths available for capital to reproduce its logic become increasingly narrow.

Among these foreseeable impacts, some are directly related to issues of imperialism. In this new world in the making, geopolitical tensions arising from pressures to control supply chains will be reinforced exponentially. The growing share of so-called renewable energy sources will increase the demand for battery production. Batteries that have, among their components, specific minerals that exist in specific deposits in specific countries. Lithium, for example, an essential mineral for battery production today, has its largest reserves in Bolivia, Chile, and Argentina.

Furthermore, the growing diplomatic pressure to draw up global agreements aimed at an energy transition capable of containing global greenhouse gas emissions has also pushed countries at the centre to seek control over mitigation opportunities. A good example is how carbon markets institutionally structure the exploitation of mitigation opportunities in peripheral countries (either on the "periphery of the centre" or on the periphery of the system).[10]

At the same time, migratory flows (both current and foreseeable) have been raising the temperature of the political environment for some years now. When we analyse the free passage for the rise of the extreme right in Europe, we cannot help but consider the tensions caused by recent migratory flows, especially from Syria. Discussing the specific case of Syria, Malm shows us that the civil war, the most apparent determinant of this migratory flow, was catalysed by a historic drought that had hit the country years before, and which could hardly be explained without reference to climate change.[11]

The disputes (economic, political, or military) over arable land, usually driven by the voracity with which the extractive logic of capital covers the planet, tend to take on much sharper contours as the first waves of impact from the climate crisis throw the most powerful nations on the globe into a quest for effective control over regions less vulnerable to these impacts.

In our contemporary world of extreme individualism due to centuries of domination and expansion of capitalist logic, perhaps it is in our favour that it is becoming increasingly difficult to hide the fact that the global ecological crisis will not spare the populations of any part of the Earth, even if some suffer sooner and more than others.

However, for alternatives to deal with the crisis we find ourselves in to flourish, we need to overcome individualistic pragmatism and move away from false alternatives in which "changes" happen just to keep everything as it is. This crisis, as the one that preceded the First World War (1914–1918) at the turn of the 19th to the 20th century, was produced by capitalist development and, like that one, cannot be overcome without radical criticism and joint action of the workers of the world.

Notes

1 See Malm, *Corona, climate, chronic emergency* (2020); Wallace, *Big farms make big flu* (2016); Wallace, *Dead epidemiologists* (2020); Foster & Suwandi, "COVID-19 and catastrophe capitalism" (2021).

2 Prado, "Questionando a macroeconomia da 'grande recessão'" ["Questioning the macroeconomics of the 'great recession'"] (2015); Foster & Suwandi, "COVID-19 and catastrophe capitalism" (2021).

3 Marx, *Capital,* vol. 1 (1982); Marx, *Capital,* vol. 2 (1992).

4 Corrêa, Sá Barreto, & Leite, "A morte lhe cai bem: capitalismo em crises e imperialismo" ["Death becomes it: capitalism in crisis and imperialism"] (2021).

5 Marx, *Capital,* vol. 1 (1982).
6 Foster & Suwandi, "COVID-19 and catastrophe capitalism" (2021), 545.
7 Foster, *Marx's ecology* (2000).
8 Sá Barreto, "Crise climática e o Green New Deal: uma primeira aproximação crítica" ["Climate crisis and the Green New Deal: a first critical approach"] (2020).
9 Sá Barreto, "Restauração neoliberal e o esgotamento histórico de formas consagradas de resistência" ["Neoliberal restoration and the historical exhaustion of established forms of resistance"] (2019).
10 Sá Barreto, "Marx contra a fantasia 'coaseana': uma crítica ontológica ao fundamento teórico dos mercados de carbono" ["Marx against the 'Coasean' fantasy: an ontological critique of the theoretical foundation of carbon markets"] (2015).
11 Malm, "Revolution in a warming world" (2017).

References

Corrêa, H., Sá Barreto, E., & Leite, L. (2021). A morte lhe cai bem: capitalismo em crises e imperialismo. *Revista Fim do Mundo, 6,* 147–180.

Foster, J. B. (2000). *Marx's ecology.* New York: Monthly Review Press.

Foster, J. B., & Suwandi, I. (2021). COVID-19 and catastrophe capitalism. In A. Callinicos, S. Kouvelakis, & L. Pradella, *Routledge handbook of Marxism and Post-Marxism.* London: Routledge.

Malm, A. (2017). Revolution in a warming world: lessons from the Russian to the Syrian revolutions. In L. Panitch, & G. Albo, *Socialist register 2017: rethinking revolution.* New York: Monthly Review Press.

Malm, A. (2020). *Corona, climate, chronic emergency: war communism in the twenty-first century.* London: Verso.

Marx, K. (1982). *Capital* (Vol. 1). London: Penguin Books.

Marx, K. (1992). *Capital* (Vol. 2). London: Penguin Books.

Prado, E. (2015). Questionando a macroeconomia da "grande recessão". *Marx e o marxismo, 3*(5), 188–220.

Sá Barreto, E. (2015). Marx contra a fantasia "coaseana": uma crítica ontológica ao fundamento teórico dos mercados de carbono. *Marx e o Marxismo, 3*(5), 263–278.

Sá Barreto, E. (2019). Restauração neoliberal e o esgotamento histórico de formas consagradas de resistência. *Revista da Sociedade Brasileira de Economia Política, 53,* 118–146.

Sá Barreto, E. (2020). Crise climática e o Green New Deal: uma primeira aproximação crítica. *Revista Fim do Mundo, 2,* 75–91.

Wallace, R. (2016). *Big farms make big flu.* New York: Monthly Review Press.

Wallace, R. (2020). *Dead epidemiologists: on the origins of COVID-19.* New York: Monthly Review Press.

Index

Note: Page numbers in **bold** denote tables.

For Product Safety Concerns and Information please contact our EU representative GPSR@taylorandfrancis.com
Taylor & Francis Verlag GmbH, Kaufingerstraße 24, 80331 München, Germany

www.ingramcontent.com/pod-product-compliance
Ingram Content Group UK Ltd.
Pitfield, Milton Keynes, MK11 3LW, UK
UKHW022354190926
475702UK00015B/2295
* 9 7 8 1 0 3 2 8 4 3 9 3 3 *